Recognition and Alleviation of Pain in Laboratory Animals

Committee on Recognition and Alleviation of Pain in Laboratory Animals

Institute for Laboratory Animal Research

Division on Earth and Life Studies

NATIONAL RESEARCH COUNCIL
OF THE NATIONAL ACADEMIES

THE NATIONAL ACADEMIES PRESS
Washington, D.C.
www.nap.edu

THE NATIONAL ACADEMIES PRESS 500 Fifth Street, NW Washington, DC 20001

NOTICE: The project that is the subject of this report was approved by the Governing Board of the National Research Council, whose members are drawn from the councils of the National Academy of Sciences, the National Academy of Engineering, and the Institute of Medicine. The members of the Committee responsible for the report were chosen for their special competences and with regard for appropriate balance.

This study was supported by the American Veterinary Medical Association; Aventis Pharmaceuticals; the Bosack-Kruger Foundation; Bristol-Myers Squibb; GlaxoSmithKline; Humane Society of the United States; Scientists Center for Animal Welfare; Wyeth-Ayerst Pharmaceuticals; Department of Agriculture, Animal and Plant Health Inspection Service; and Department of Health and Human Services, National Institutes of Health through Contract Number N01-OD-4-2139 Task Order #161. Any opinions, findings, conclusions, or recommendations expressed in this publication are those of the authors and do not necessarily reflect the views of the organizations or agencies that provided support for the project. The content of this publication does not necessarily reflect the views or policies of the National Institutes of Health, nor does mention of trade names, commercial products, or organizations imply endorsement by the US government.

Library of Congress Cataloging-in-Publication Data

Institute for Laboratory Animal Research (U.S.). Committee on Recognition and Alleviation of Distress in Laboratory Animals.
Recognition and alleviation of pain in laboratory animals / Committee on Recognition and Alleviation of Pain in Laboratory Animals, Institute for Laboratory Animal Research, Division on Earth and Life Studies, National Research Council of the National Academies.
p. cm.
Includes bibliographical references and index.
ISBN-13: 978-0-309-12834-6 (pbk.)
ISBN-10: 0-309-12834-X (pbk.)
ISBN-13: 978-0-309-12835-3 (pdf)
ISBN-10: 0-309-12835-8 (pdf)
1. Laboratory animals—Health. 2. Laboratory animals—Effect of stress on. 3. Pain in animals—Treatment. 4. Animal welfare. I. Title.
SF406.I565 2009
636.088′5—dc22

2009042189

Additional copies of this report are available from The National Academies Press, 500 Fifth Street, NW, Lockbox 285, Washington, DC 20001; (800) 624-6242 or (202) 334-3313 (in the Washington metropolitan area); Internet, http://www.nap.edu

Printed in the United States of America

THE NATIONAL ACADEMIES

Advisers to the Nation on Science, Engineering, and Medicine

The **National Academy of Sciences** is a private, nonprofit, self-perpetuating society of distinguished scholars engaged in scientific and engineering research, dedicated to the furtherance of science and technology and to their use for the general welfare. Upon the authority of the charter granted to it by the Congress in 1863, the Academy has a mandate that requires it to advise the federal government on scientific and technical matters. Dr. Ralph J. Cicerone is president of the National Academy of Sciences.

The **National Academy of Engineering** was established in 1964, under the charter of the National Academy of Sciences, as a parallel organization of outstanding engineers. It is autonomous in its administration and in the selection of its members, sharing with the National Academy of Sciences the responsibility for advising the federal government. The National Academy of Engineering also sponsors engineering programs aimed at meeting national needs, encourages education and research, and recognizes the superior achievements of engineers. Dr. Charles M. Vest is president of the National Academy of Engineering.

The **Institute of Medicine** was established in 1970 by the National Academy of Sciences to secure the services of eminent members of appropriate professions in the examination of policy matters pertaining to the health of the public. The Institute acts under the responsibility given to the National Academy of Sciences by its congressional charter to be an adviser to the federal government and, upon its own initiative, to identify issues of medical care, research, and education. Dr. Harvey V. Fineberg is president of the Institute of Medicine.

The **National Research Council** was organized by the National Academy of Sciences in 1916 to associate the broad community of science and technology with the Academy's purposes of furthering knowledge and advising the federal government. Functioning in accordance with general policies determined by the Academy, the Council has become the principal operating agency of both the National Academy of Sciences and the National Academy of Engineering in providing services to the government, the public, and the scientific and engineering communities. The Council is administered jointly by both Academies and the Institute of Medicine. Dr. Ralph J. Cicerone and Dr. Charles M. Vest are chair and vice chair, respectively, of the National Research Council.

www.national-academies.org

COMMITTEE ON RECOGNITION AND ALLEVIATION OF PAIN IN LABORATORY ANIMALS

Gerald F. Gebhart (*Chair*), University of Pittsburgh
Allan I. Basbaum, University of California
Stephanie J. Bird, Waltham, Massachusetts
Paul Flecknell, Newcastle University, United Kingdom
Lyndon Goodly, University of Illinois
Alicia Z. Karas, Tufts University
Stephen T. Kelley, University of Washington
Jane Lacher, Dow Chemical Company
Georgia Mason, University of Guelph, Canada
Lynne U. Sneddon, University of Liverpool, United Kingdom
Sulpicio G. Soriano, Harvard University

Consultant

Heidi L. Shafford, Veterinary Anesthesia Specialists, LLC

Staff

Joanne Zurlo, Director
Lida Anestidou, Study Director
Kathleen Beil, Administrative Coordinator
Cameron H. Fletcher, Senior Editor
Rhonda Haycraft, Senior Project Assistant

INSTITUTE FOR LABORATORY ANIMAL RESEARCH COUNCIL

INSTITUTE FOR LABORATORY ANIMAL RESEARCH PUBLICATIONS

Scientific and Humane Issues in the Use of Random Source Dogs and Cats in Research (2009)
Recognition and Alleviation of Distress in Laboratory Animals (2008)
Toxicity Testing in the 21st Century: A Vision and a Strategy (2007)
Overcoming Challenges to Develop Countermeasures Against Aerosolized Bioterrorism Agents: Appropriate Use of Animal Models (2006)
Guidelines for the Humane Transportation of Research Animals (2006)
Science, Medicine, and Animals: Teacher's Guide (2005)
Animal Care and Management at the National Zoo: Final Report (2005)
Science, Medicine, and Animals (2004)
The Development of Science-based Guidelines for Laboratory Animal Care: Proceedings of the November 2003 International Workshop (2004)
Animal Care and Management at the National Zoo: Interim Report (2004)
National Need and Priorities for Veterinarians in Biomedical Research (2004)
Guidelines for the Care and Use of Mammals in Neuroscience and Behavioral Research (2003)
International Perspectives: The Future of Nonhuman Primate Resources, Proceedings of the Workshop Held April 17-19, 2002 (2003)
Occupational Health and Safety in the Care and Use of Nonhuman Primates (2003)
Definition of Pain and Distress and Reporting Requirements for Laboratory Animals: Proceedings of the Workshop Held June 22, 2000 (2000)
Strategies That Influence Cost Containment in Animal Research Facilities (2000)
Microbial Status and Genetic Evaluation of Mice and Rats: Proceedings of the 1999 US/Japan Conference (2000)
Microbial and Phenotypic Definition of Rats and Mice: Proceedings of the 1998 US/Japan Conference (1999)
Monoclonal Antibody Production (1999)
The Psychological Well-Being of Nonhuman Primates (1998)
Biomedical Models and Resources: Current Needs and Future Opportunities (1998)
Approaches to Cost Recovery for Animal Research: Implications for Science, Animals, Research Competitiveness and Regulatory Compliance (1998)
Chimpanzees in Research: Strategies for Their Ethical Care, Management, and Use (1997)
Occupational Health and Safety in the Care and Use of Research Animals (1997)

Guide for the Care and Use of Laboratory Animals (1996)
Guide for the Care and Use of Laboratory Animals translations: Korean, Chinese, Spanish, Russian, French, Taiwanese, Portugese, Japanese, Arabic, Turkish (1996)
Rodents (1996)
Nutrient Requirements of Laboratory Animals, Fourth Revised Edition (1995)
Laboratory Animal Management: Dogs (1994)
Recognition and Alleviation of Pain and Distress in Laboratory Animals (1992)
Education and Training in the Care and Use of Laboratory Animals: A Guide for Developing Institutional Programs (1991)
Companion Guide to Infectious Diseases of Mice and Rats (1991)
Infectious Diseases of Mice and Rats (1991)
Immunodeficient Rodents: A Guide to Their Immunobiology, Husbandry, and Use (1989)
Use of Laboratory Animals in Biomedical and Behavioral Research (1988)
Animals for Research: A Directory of Sources, Tenth Edition and Supplement (1979)
Amphibians: Guidelines for the Breeding, Care and Management of Laboratory Animals (1974)

Copies of these reports can be ordered from the National Academies Press
(800) 624-6242 or (202) 334-3313
www.nap.edu

Acknowledgments

In preparation for this report, the committee invited experts to present their perspectives on the concepts of nociception, pain, consciousness, and awareness. The committee thanks:

Colin Allen, Indiana University
A. Vania Apkarian, Northwestern University
David Borsook, McLean Hospital

This report was reviewed in draft form by individuals chosen for their diverse perspectives and technical expertise, in accordance with procedures approved by the Report Review Committee of the National Research Council (NRC). The purpose of this independent review is to provide candid and critical comments that will assist the committee in making its published report as sound as possible and to ensure that the report meets institutional standards for objectivity, evidence, and responsiveness to the study charge. The review comments and draft manuscript remain confidential to protect the integrity of the deliberation process. The committee thanks the following individuals for their review of this report:

K. S. Anand, University of Arkansas for Medical Sciences
George J. DeMarco, Pfizer, Inc.
Ronald Dubner, University of Maryland
Sherril Green, Stanford School of Medicine
C. Terrance Hawk, GlaxoSmithKline Pharmaceuticals R&D

B. Duncan X. Lascelles, North Carolina State College of Veterinary Medicine
Jerald Silverman, University of Massachusetts Medical School
William S. Stokes, National Institute of Environmental and Health Sciences
Daniel M. Weary, University of British Columbia, Canada
Tony L. Yaksh, University of California—San Diego

Although the reviewers listed above provided many constructive comments and suggestions, they were not asked to endorse the report's conclusions or recommendations, nor did they see the final draft of the report before its release. The review of this report was overseen by **Hilton J. Klein**, Taconic, and **Harley W. Moon**, Iowa State University (*emeritus*). Appointed by the NRC, they were responsible for making certain that an independent examination of this report was carried out in accordance with institutional procedures and that all review comments were carefully considered. Responsibility for the final content of this report rests entirely with the authoring committee and the institution.

This report is the product of committee members who gave generously of their time and effort. The committee rapidly developed into a collegial, hard-working group, freely shared ideas, debated contentious issues enthusiastically, and strived to make this report both useful and informative to readers. We grew in the process and learned from each other. I deeply appreciate the members' contributions and insistence on applying an evidence-based approach to the content and recommendations in the report. Their efforts would not have been successful without the invaluable help of ILAR staff and committee consultants, particularly Lida Anestidou and Heidi Shafford, respectively, who each deserve our sincere thanks. I am deeply appreciative of the opportunity to have been a part of this effort and anticipate that the report will meet its principal objectives.

Gerald F. Gebhart, *Chair*
Committee on Recognition and Alleviation
of Pain in Laboratory Animals

Glossary

Affect: The positive (i.e., preferred) and negative (i.e., avoided) states experienced by animals. Affect is a conscious experience (see consciousness). It is similar to the colloquial use of the term "emotion."

Allodynia: Pain produced by normally nonnoxious stimuli (e.g., touch).

Analgesic: A drug or endogenous mediator that relieves/reduces pain without concomitant loss of consciousness (e.g., morphine). However, opioid analgesics, as well as most drugs used to relieve pain, have sedative-hypnotic properties at greater doses.

Anesthetic: A drug that eliminates sensation, including the experience of pain; depending on its activity, it may or may not eliminate pain by inducing loss of consciousness (e.g., local anesthetic vs. barbiturate).

Animal welfare: In this report we use "welfare" to mean "well-being."

Anxiolytics: Drugs that reduce anxiety, often used in combination with other drugs to manage pain.

Awareness: Feeling, or the experienced state that accompanies pain and other sensations (and thus distinguishes pain from nociception). This report uses "awareness" and "consciousness" interchangeably.

Central sensitization: Increased excitability of central nervous system

(CNS) neurons and consequent amplification of input initiated by sensitized nociceptors.

Consciousness: This term has a range of meanings; in this report it refers to the experience of sensation widely shared by most animals.

Hyperalgesia: Increased sensitivity and response to a noxious stimulus enhanced by sensitization of peripheral nociceptors and central neurons (opposite is hypoalgesia).

Inappetence: Lack of appetite.

Neuraxis: The central nervous system (CNS; the spinal cord and the brain).

Nociception: The detection of a noxious event by nociceptors. Nociception represents the peripheral and central nervous system processing of information about the internal or external environment generated by nociceptor activation.

Nociceptor sensitization: Increased excitability and response of nociceptors produced by endogenous mediators (e.g., prostaglandins, protons).

Noxious stimulus and nociceptors: An event that damages or threatens to damage tissues and that activates specialized sensory nerve endings called nociceptors.

Operant conditioning: The use of positive and negative consequences to modify behavior through learning.

Pain: An unpleasant sensory and emotional experience associated with actual or potential tissue damage, or described in terms of such damage.

Pain descriptors

1. *Momentary pain:* short-lasting, brief, transient (e.g., seconds) and usually of low intensity.
2. *Postprocedural/postsurgical pain:* longer-lasting than momentary (hours to days to weeks), a consequence of tissue injury due to surgery or other procedures.
3. *Persistent pain:* lasts for days to weeks such as encountered in studies that investigate pain (and caused by mechanisms other than postprocedural pain).
4. *Chronic pain:* pain of long duration (i.e., days to weeks to months), typically associated with degenerative diseases, without relief, difficult to manage clinically.

Contents

Summary

This report is the response to a request by the New Jersey Association for Biomedical Research that the Institute for Laboratory Animal Research form a consensus committee to update the 1992 National Research Council (NRC) report *Recognition and Alleviation of Pain and Distress in Laboratory Animals*. In the 16 years since the first report was published, there has been significant scientific progress in the areas of animal welfare, stress, distress, and pain to warrant a fresh look at the topics of that report. This report follows the release of the 2008 NRC report *Recognition and Alleviation of Distress in Laboratory Animals*.

Although the numerous regulations, policies, and guidelines that govern animal use in research in the United States address distress and pain jointly, from a scientific perspective the two concepts are quite distinct. According to the International Association for the Study of Pain, pain in humans is "an unpleasant sensory and emotional experience associated with actual or potential tissue damage, or described in terms of such damage" (IASP 1979). Pain is mediated through the activity of specialized sensory receptors, called nociceptors, involves the possibility of bodily injury, and depends on the interaction between those nociceptors and higher processing centers in the brain to generate the negative emotional component associated with the potential harm. While pain can be detrimental to animal welfare, distress always is, as it is a measure of the animal's inability to cope with a stressor.

Adopting an approach similar to that of the 2008 report, the committee that prepared this report focused on the management and avoidance of pain wherever scientifically possible. Continuing in the steps of the 1992 committee, the current committee embraced the idea that in most experimental and husbandry situations laboratory animals need not experience pain, and that its alleviation and prevention are an ethical and moral imperative that is embodied in the relevant regulations and policies. In fact, this approach was codified in the statement of task for this project:

> The . . . report will update information based on the current scientific literature on recognizing and alleviating pain in laboratory animals. The report will discuss the physiology of pain in commonly used laboratory species. Specific emphasis will be placed on the identification of humane endpoints, pharmacologic and non-pharmacologic principles to control pain, and principles to utilize in minimizing pain associated with experimental procedures. As with the first report [on Distress], general guidelines and examples will be given to aid IACUC members, investigators and animal care staff in making decisions about protocols using laboratory animals under current federal regulations and policies.

APPROACH TO THIS STUDY

The committee collected and evaluated scientific evidence from peer-reviewed published literature, evidence-based veterinary practices, and expert opinions, and defined a consistent terminology in the Glossary and Chapter 1. The committee examined the occurrence of pain in vertebrates alone, for several reasons: (1) the current regulations affect only the vertebrate phylum; (2) most laboratory animal species used in research, education, and training are vertebrates; and (3) there is ongoing debate about whether pain occurs in subjects that may or may not have consciousness (readers are urged to explore studies of adult humans in a persistent vegetative state or with dementia and consider the implications of those data for nonverbal populations such as laboratory animals). As it was beyond the task of this committee to evaluate and analyze the last question, the underlying premise of this report is that **all vertebrates should be considered capable of experiencing the aversive state of pain**.

Although most of the information used in the report reflects studies and observations in mammals, currently available (albeit very limited) data on birds, fish, reptiles, and amphibians are also included. The committee decided against including information on the treatment and management of pain for each laboratory species, because for the commonest of these many referenced and peer-reviewed publications, professional societies' guidelines, books, and book chapters are readily available for reference. Instead, the committee opted to expand on species for which the body of peer-reviewed work is still small and for which guidelines are lacking. This report therefore provides practical information on birds, amphibians, fish, and reptiles in order to help the scientific and veterinary community better care for these laboratory species.

PAIN IN ANIMAL RESEARCH

The committee acknowledges that pain in animals is difficult to assess, mostly because of a lack of methods to validate and objectively measure it. Until such tools are developed, behavioral indices and careful extrapolation from the human experience should be used to assess pain in research animals. It is important to bear in mind that pain may be not only the result of a research procedure but also a byproduct of husbandry or other unrelated factors (e.g., aging). Pain may arise in response to a noxious stimulus and in situations likely to cause increased sensitivity to pain (i.e., hyperalgesia), such as injury and inflammation. Psychological factors also likely contribute to pain under these circumstances.

Pain is the result of a cascade of physiological, immunological, cognitive, and behavioral effects that may make uncontrolled pain a source of experimental error. Although there are circumstances in which withholding treatment is necessary (as, for example, when pain itself is the focus of the study), routinely withholding analgesics after surgery or other invasive procedures with anticipated moderate to severe pain is detrimental to the welfare of the research subjects, contrary to the regulatory mandate, and unethical. A useful assumption is that the magnitude of the clinical signs (see Chapter 3) and behavioral changes observed correlates closely with the intensity of pain. Current best practices to assess pain entail a structured clinical examination combined with solid knowledge of the normal appearance and behavior of the species.

Anticipating the potential intensity of pain is important in designing the most appropriate approach to its management or prevention. Common interventions to treat pain include the use of anesthetics, analgesics, anxiolytics, and nonpharmacological methods. Although regulations specify that only nonbrief, procedural pain requires treatment, pain of any duration or intensity—including multiple episodes of momentary pain—merits consideration and potential treatment.

As in *Recognition and Alleviation of Distress in Laboratory Animals,* this report stresses the importance of the *Three Rs* (replacement, refinement, and reduction) as the standard for identifying, modifying, minimizing, and avoiding most causes of pain in laboratory animals. To this end, the committee believes that adoption of humane endpoints is critical, particularly in studies where significant pain is anticipated. Because humane endpoints are unique to individual research projects, pilot studies should be undertaken to identify and incorporate them in the study design. In this as in all stages of the research, good communication between researchers, veterinarians, animal care personnel, and institutional animal care and use committee (IACUC) members is crucial.

RECOMMENDATIONS

Based on the information analyzed and discussed in this report, the committee makes overarching consensus recommendations above and beyond those at the end of individual chapters:

- Current scientific evidence strongly suggests that mammals, including rodents (the most commonly used laboratory animals), are able to experience pain. Researchers, veterinarians, animal care personnel, and IACUC members should heed the 4th Government Principle[1] and use professional judgment and best practices to avoid or minimize unnecessary pain. Researchers conducting studies in which more than momentary pain is anticipated should, in addition to providing appropriate analgesia, consider and enforce (where possible) humane endpoints to protect the welfare of the laboratory animals.
- Knowledge about pain in nonmammalian species is incomplete and in the absence of evidence they should be treated humanely with serious consideration of, and attention to, the potentially painful implications of noxious stimuli and invasive procedures.
- Any study that will likely result in pain for the animal subjects should have clearly determined, appropriate humane endpoints. The importance of pilot studies to determine such endpoints is paramount. Teamwork and open communication between researchers, veterinarians, animal care staff, and the IACUC can facilitate and expedite the definition, validation, and implementation of appropriate endpoints.
- Funding is particularly difficult for projects that investigate the understanding, recognition, and alleviation of pain, especially if the beneficiaries of such studies are the laboratory animals themselves. However, lack of knowledge of drug effects and doses in many mammalian and especially nonmammalian species, and the potentially confounding effects of analgesics and anesthetics on study variables, limit effective pain management. Given the impact of better animal welfare on science as well as the growing public interest in the treatment of laboratory animals, federal agencies and foundations that support biomedical and behavioral research should make funds available for pain-related studies (see also NRC 2008).

[1] US Government Principle #4 states that "Proper use of animals, including the avoidance or minimization of discomfort, distress, and pain when consistent with sound scientific practices, is imperative. Unless the contrary is established, investigators should consider that procedures that cause pain or distress in human beings may cause pain or distress in other animals."

- Lack of adequate funding also hinders efforts to develop and validate alternatives (methods, procedures, and research strategies); such efforts must continue in order to ensure the incorporation of alternatives in research projects and safety assessment tests.
- It is necessary to educate investigators, veterinarians, and animal care staff about the basic physiologic principles, causes, signs, and availability of diverse treatment options and potential deleterious effects of those treatments on pain. As the field of pain medicine benefits from new insights and methods of prevention and treatment for humans, so should laboratory animals benefit from the research for which they are a currently indispensable underpinning. As laboratory animal veterinarians enhance their understanding of pain management and regulatory policy is updated, the ability to minimize pain in laboratory animals can proceed in tandem with scientific progress.

REFERENCES

IASP (International Association for the Study of Pain). 1979. IASP Pain Terminology. Available at http://www.iasp-pain.org/AM/Template.cfm?Section=Pain_Definitions&Template=/CM/HTMLDisplay.cfm&ContentID=1728#Pain. Accessed January 8, 2009.

NRC (National Research Council). 2008. Recognition and Alleviation of Distress in Laboratory Animals. Washington: National Academies Press.

Introduction

In 1992, the National Research Council published a report titled *Recognition and Alleviation of Pain and Distress in Laboratory Animals* "to help scientists, research administrators, institutional animal care and use committees (IACUCs), and animal care staff to address the difficult questions of the presence and alleviation of animal pain and distress" (NRC 1992, p. 1). The need for assistance in this area has persisted and, with new scientific discoveries, the generation of genetically modified animals, and continued regulatory emphasis on minimizing pain and distress in laboratory animals, it became evident that the 1992 report had become outdated. The Institute for Laboratory Animal Research received several requests from the veterinary and biomedical communities to convene a committee to update the report. After many discussions with constituents and several sponsors, the National Academies opted to update the 1992 report as two separate reports, one on distress and one on pain, because although they are linked in regulation, they are quite different scientifically (NRC 2008).

This report on the *Recognition and Alleviation of Pain in Laboratory Animals* was prepared to help scientists, veterinarians, research administrators, IACUCs, and animal care staff understand the basis of animal pain, recognize and evaluate its presence and severity, and appreciate means to minimize or abolish pain, according to the charge to the committee that prepared this report:

> The . . . report will update information based on the current scientific literature on recognizing and alleviating pain in laboratory animals. The report will discuss the physiology of pain in commonly used laboratory species.

> Specific emphasis will be placed on the identification of humane endpoints, pharmacologic and non-pharmacologic principles to control pain, and principles to utilize in minimizing pain associated with experimental procedures. As with the first report [on Distress], general guidelines and examples will be given to aid IACUC members, investigators and animal care staff in making decisions about protocols using laboratory animals under current federal regulations and policies.

The committee believes that in most experimental and husbandry situations laboratory animals need not experience ongoing or substantial pain and that prevention and alleviation of pain in laboratory animals is an ethical imperative. This view, shared by the public and Congress as well as federal agencies and organizations, is codified in laws, regulations, policies, recommendations, and guidelines (presented in Appendix B) that govern the care and use of animals in research and that require the identification, minimization, and elimination of sources of pain, unless the scientific merit of a study demands otherwise. These regulations, policies, and guidelines also require that institutions develop programs for training personnel in procedures to ensure the minimization of animal pain.

The purposes of this report are to increase awareness of the sources and recognition of pain in laboratory animals and to increase ethical sensitivity in those who use and care for them. The report may also, indirectly, help to reduce the number of animals needed for experimental purposes because uncontrolled pain can increase variability in experimental data and so require the use of more animals. If this report improves investigators' awareness of their obligations for the humane care and use of their research animals, it could even reduce the replication required to establish the generality of their scientific findings. Such a reduction, however, should always be consistent with the necessity to validate important scientific findings.

ORGANIZATION OF THE REPORT

This report focuses on the principles of recognizing pain and on pharmacologic and nonpharmacologic methods of minimizing and controlling pain. It was not planned as a source of information on experimental design, nor was it designed as a training document, although it may certainly be useful for this purpose (the report *Education and Training in the Care and Use of Laboratory Animals* might be of more direct assistance with the development of training and education programs; NRC 1991).

Chapters 2 and 3 focus on what is known about the biology and physiology of pain and how to recognize and assess it in animals. Chapters 4 and 5, respectively, provide information about controlling pain, with species-specific recommendations, and humane endpoints. Appendix A provides information on pain as a study subject, and Appendix B lists the regulatory and legal requirements that apply to pain recognition and management in the use of animals in research.

The intent of this report is to help veterinarians, investigators, researchers, IACUC members, and animal care staff understand pain in order to adequately manage and if possible avoid it. The committee compiled the most up-to-date information available but also relied on its scientific expertise to make recommendations to uphold the principles of humane care and use of laboratory animals. The committee urges readers to consider this information carefully and hopes that this report will help link the integrity of scientific methodology to the humane care of animal subjects.

REFERENCES

NRC (National Research Council). 1991. Education and Training in the Care and Use of Laboratory Animals. Washington: National Academy Press.

NRC. 1992. Recognition and Alleviation of Pain and Distress in Laboratory Animals. Washington: National Academy Press.

NRC. 2008. Recognition and Alleviation of Distress in Laboratory Animals. Washington: National Academies Press.

1

Pain in Research Animals: General Principles and Considerations

This chapter presents an overview of the ethical, legal, and scientific reasons that mandate the alleviation of animal pain, drawing attention to the principles of the *Three Rs* (3Rs; replacement, refinement, and reduction) and the central role of refinement in the humane care and use of laboratory animals. It includes discussion of the fundamental concepts of the experience of pain and factors that affect pain aversiveness. It focuses on the potential causes of pain in research animals while broadly considering evidence of pain in vertebrates. It concludes with a discussion of the particular circumstances that may justify pain in laboratory animals.

WHY IS IT IMPORTANT TO RECOGNIZE AND ALLEVIATE ANIMAL PAIN?

Most research using animals is for the direct or indirect benefit of society. Furthermore, most research on animals is funded, directly or indirectly, by the public. For both these reasons, the public has the right and responsibility to discuss how animal research is conducted. The public expects animal experimentation to be not only scientifically justifiable and valid but also humane, meaning that it results in minimal or no pain, stress, distress, or other negative impact on the welfare of the animals involved. When laboratory animals are subjected to conditions that do cause pain or distress, then ethically—at least from a utilitarian perspective—the benefits must outweigh the costs. This ethical justification depends on the challenging balance between the benefits (primarily to humans) and the costs to experimental animals in the form of pain, distress, and euthanasia.

These ethical expectations are embodied in the principles of the *Three Rs*: replacement, refinement, and reduction (the 3Rs; Russell and Burch 1959). As outlined in Appendix B, they are also enforced by laws and encouraged by professional guidelines. The 3Rs, formulated to protect the welfare of animals used in research, are widely accepted as international standards for the humane use of animals in research or testing. The National Centre for the Replacement, Refinement and Reduction of Animals in Research (NC3Rs; http://www.nc3rs.org.uk) defines the *Three Rs* as follows:

- *Replacement* refers to methods that replace or avoid the use of animals. Examples include the use of alternative methods (e.g., computer modeling, in vitro methods) or the replacement of higher-order animals such as mammals with "lower" animals (e.g., invertebrates, such as *Drosophila* and nematode worms).
- *Refinement* refers to improvements to animal welfare in studies where the use of animals is unavoidable. Such improvements affect the lifetime experience of the animal and apply to husbandry or procedures that improve welfare and/or minimize pain, distress, lasting harm, or other threats to welfare. Examples of refinement include training animals to cooperate with certain procedures (e.g., blood sampling) to reduce stress, ensuring that accommodation meets animals' needs (e.g., socially housing primates), and using appropriate anesthetic and analgesic drugs. The committee also urges the definition of humane endpoints for each experiment as an important refinement.
- *Reduction* refers to methods that minimize animal use and enable researchers to obtain equivalent information from fewer animals or more information from the same number of animals. Such methods include appropriate experimental design, sample size determination, statistical analysis, and the use of advanced noninvasive imaging techniques.

The principle of refinement, especially in the context of animal pain, is central to many US regulations and guidelines (see Appendix B): almost all specify that procedures involving animals should (1) avoid or minimize discomfort and pain, and/or (2) otherwise include the provision of adequate pain relief unless the pain is justified scientifically.

Minimizing animal pain whenever possible is thus important both ethically and legally. It is also a practice that yields scientific and practical benefits, as discussed in Chapters 2 and 4. For example, the early experience of pain in postnatal animals may lead to increased pain sensitivity in the insulted tissue later in life (Chapter 2), while effective pain management in all animals (Chapter 4) may improve healing rates, decrease mortality, and prevent the potentially confounding effects of untreated pain on many aspects of biological function (e.g., immune function, sleep, cognition, and many biological variables that are affected by stress; for discussion see Chapter 2).

WHAT IS PAIN?

Essential to any discussion of how to avoid or minimize pain in animals is a clear understanding and definition of pain and related terms. What exactly is pain? How does it differ from "nociception"? How does pain vary? And what dimensions of pain are most relevant to animal welfare?

The International Association for the Study of Pain (IASP; www.iasp-pain.org) defines pain in humans as "an unpleasant sensory and emotional experience associated with actual or potential tissue damage, or described in terms of such damage" (IASP 1979). Pain typically involves a noxious stimulus or event that activates nociceptors in the body's tissues that convey signals to the central nervous system, where they are processed and generate multiple responses, including the "unpleasant sensory and emotional experience" central to the IASP definition. The anatomy and biology of pain are covered in more detail in Chapter 2. Some key issues and important terms are addressed below to highlight some of the challenges in understanding animal pain.

Noxious Stimuli and Nociception

"Noxious stimuli" are events that damage or threaten damage to tissues (e.g., cutting, crushing, or burning stimuli) and that activate specialized sensory nerve endings called nociceptors. First described in the skin by Sherrington in 1906, nociceptors are also in muscle, joints, and viscera. Sherrington coined the term "nociception" to describe the detection of a noxious event by nociceptors. Nociception thus represents the peripheral and central nervous system processing of information about the internal or external environment as generated by nociceptor activation. This information is processed at both spinal and supraspinal levels of the central nervous system, providing details about the nature, intensity, location, and duration of noxious events.

It is important to understand that stimuli adequate to activate nociceptors are not the same for all tissues; following are examples of common types of noxious stimuli for different tissues:

- Skin: thermal (hot or cold), mechanical (cutting, pinching, crushing), and chemical (inflammatory and other mediators released from or synthesized by damaged skin, and exogenous chemical stimuli such as formalin, carrageenan, bee venom, capsaicin)

- Joints: mechanical (rotation/torque beyond the joint's normal range of motion) and chemical (inflammatory and other mediators released into or injected into the joint capsule)
- Muscle: mechanical (blunt force, stretching, crushing, overuse) and chemical (inflammatory and other mediators released from or injected into muscle)
- Viscera: mechanical (distension, traction on the mesentery) and chemical (inflammatory and other mediators released from inflamed or ischemic organs, inhaled irritants).

Noxious stimulation triggers multiple physiological and behavioral responses, only one of which is the generation of the unpleasant emotional state of pain. Other behavioral and physiological responses include withdrawal reflexes, increases in heart rate and blood pressure, and other parameters. As discussed below (see Boxes 1-3 and 1-4), many of these responses can also occur in organisms that do *not* experience pain (e.g., anesthetized animals, or those with spinal lesions that prevent nociceptive information from reaching higher central nervous system structures). Thus pain and nociception are distinct concepts, and some nociceptive responses (e.g., withdrawal reflexes in spinal cord-transected animals) do not necessarily indicate pain. However, in the intact animal and in humans, nociceptive input reaches subcortical and cortical brain nuclei that contribute to the affective, aversive states of pain. In humans, therefore, nociceptive reflex withdrawal responses generally correlate with experiences of pain as evidenced by verbal feedback about the quality of the stimulus. Nonhuman animals cannot provide verbal feedback. Therefore, an ongoing challenge in laboratory animal research is to determine whether responses that could merely be nociceptive are also indicative of pain, and, conversely, whether the abolition of nociceptive responses indicates the successful abolition of pain. Thus, in the intact animal (e.g., under light anesthesia that removes some but not all responses to noxious stimuli), the distinction between nociception and pain is not always clear.

Pain

The generation of pain from nociceptive signals occurs in the central nervous system (CNS). Certain regions of the forebrain are responsible for the experience of both the sensory aspects of pain (i.e., qualitative properties such as location, duration, and whether "sharp" or "dull") and the unpleasant, affective aspects associated with it (i.e., the way that pain "hurts"; Baliki et al. 2006; for details see Chapter 2). Studies of human pain have shown that pain is unpleasant and aversive: humans typically seek to avoid and minimize it. Furthermore, anticipation or threats of pain can cause anxiety and/or fear (Price 2002). This so-called "negative valence" of pain (i.e., the fact that it is aversive) underlies its description as emotional/affective (Box 1-2).

Aversiveness is thus a consistent characteristic of pain, but does not mean that all pain is the same: it varies in character (e.g., stinging, throbbing, aching, burning), location (e.g., joints, viscera), duration (from momentary to persistent or chronic), and intensity (from minimal to very intense). Pain can thus vary in its sensory, qualitative properties as well as in the extent of its aversiveness or unpleasantness. How aversive or unpleasant pain is depends primarily on its duration and intensity (Price 2002), although as explained below, psychological factors such as controllability can also affect the experience of pain.

In terms of duration momentary pain is less aversive than persistent or chronic pain (see Box 1-1 for terminology and definitions). Indeed, many animals (and humans) are prepared to accept momentary discomfort or pain (e.g., that from a needle stick) especially if it is associated with a reward. In contrast, chronic pain (e.g., that caused by osteoarthritis or cancer) can be very difficult to manage and thus lead to distress and pathological changes that further undermine well-being (e.g., hypertension, immunosuppression, depression, cognitive changes, and possibly structural changes in the brain; Apkarian et al. 2004a,b).

Similarly, intensity affects the aversiveness of pain. Intensity can vary from very low, when pain is first detected (the "pain threshold"), to the upper limit of tolerance and beyond (where tolerance is defined as the greatest intensity of pain that is accepted voluntarily). Obviously, more intense, severe pain is more aversive than slight pain.

In humans, physiological and/or psychological state (e.g., stress, anxiety, fear) can also alter the aversiveness of pain (Carlsson et al. 2006; Keogh and Cochrane 2002; Price 2002). For example, pain that is controllable, predictable, or seen as ultimately yielding some benefit (e.g., the birth of a much-wanted child) is typically reported by humans as more tolerable and less aversive than uncontrollable, unpredictable pain of the same quality and intensity.

Emerging evidence suggests that this may be true for some laboratory animals as well (Gentle 2001; Langford et al. 2006). Such factors are, however, far less well understood for animals. Thus, efforts to alleviate pain in research animals typically focus on reducing its duration and/or intensity. Figure 1-1 helps illustrate how duration and intensity interact to affect aversiveness. Indeed, the phrase "more than *momentary* or *slight* pain" appears repeatedly in animal protection legislation and guidelines[1] (see Appendix B) to emphasize that longer-lasting or more intense pain should cause ethi-

[1]For example, the duration and intensity of pain are central to USDA animal pain categories (where C refers to "minimal, transient, or no pain or distress," and D and E procedures refer to "more than minimal or transient pain/or distress"; USDA 1997a).

BOX 1-1
Terms Referring to the Duration of Pain

The variety of terms used to describe the duration of pain can be imprecise and confusing, particularly because clinicians (e.g., veterinarians) and pain researchers differ in their vocabulary. We present the terms here and explain how they are used in this report.

Acute pain is used by pain researchers to refer to pain that is momentary, such as associated with a needle stick (e.g., drug injection, venipuncture) or an experimentally applied noxious stimulus that does not produce noticeable tissue damage (e.g., pinch, mild electric shock). These experimental manipulations may generate a withdrawal reflex or vocalization. However, this pain is of very short duration (seconds to tens of seconds, perhaps minutes when assessing pain tolerance; see further discussion in text) and consequences to the subject are minimal and brief. In this document, **momentary pain** refers to this kind of brief, transient pain.

Acute pain is also used in both human and animal clinical medicine to label the pain typically associated with procedures or surgery. Tissue injury is a usual consequence of such procedures and thus the pain induced is considerably longer lasting than momentary (e.g., lasting for days to more than a week). In this document, pain of this nature is referred to as **postprocedural** or **postsurgical pain**.[a]

Persistent pain refers in this report to pain states that can last for days or weeks but that are caused by different mechanisms than momentary or postprocedural pain.[b] To study these mechanisms numerous animal models have been developed that are commonly known as "persistent pain models." These are described in Appendix A.

Chronic pain, of long duration (weeks, months, or years), can be difficult to manage in both human and animal clinical settings.[c] These pain states are distinct from postprocedural or persistent pain in that they are typically associated with tissue degenerative and destructive diseases (e.g., osteoarthritis, cancer) and do not improve or resolve over time. In the context of laboratory animal medicine, chronic pain is most commonly a byproduct of non-pain-related research (e.g., aging, disease research).

[a]The committee recognizes that the term "acute pain" is commonly used by human and animal clinicians/veterinarians to refer to postprocedural pain or "sharp" pain. However, "sharp" pain can be of both short or long duration (usually undefined), and "acute" means different things to different people. The committee therefore abstains from using the terms "acute" or "sharp" in favor of the terms "momentary" and "postprocedural" or "postsurgical" as defined above.

[b]A common synonym for "persistent" is "tonic," a description commonly used in pain research, that characterizes pain evoked for as long as nociceptors are stimulated.

[c]Chronic pain in humans is usually defined as pain lasting beyond the expected course of normal healing, often arbitrarily set at 6 months or beyond. Such duration is not appropriate to apply to laboratory animals with much shorter lifespans than humans or early developmental stages. Recurring or constant pain that lasts beyond the expected course of normal healing (which differs per species and per insult/injury) may merit consideration as "chronic pain." The committee urges pain researchers, veterinarians, animal care staff, and IACUCs to recognize the influence of lifespan on the definition of chronic pain.

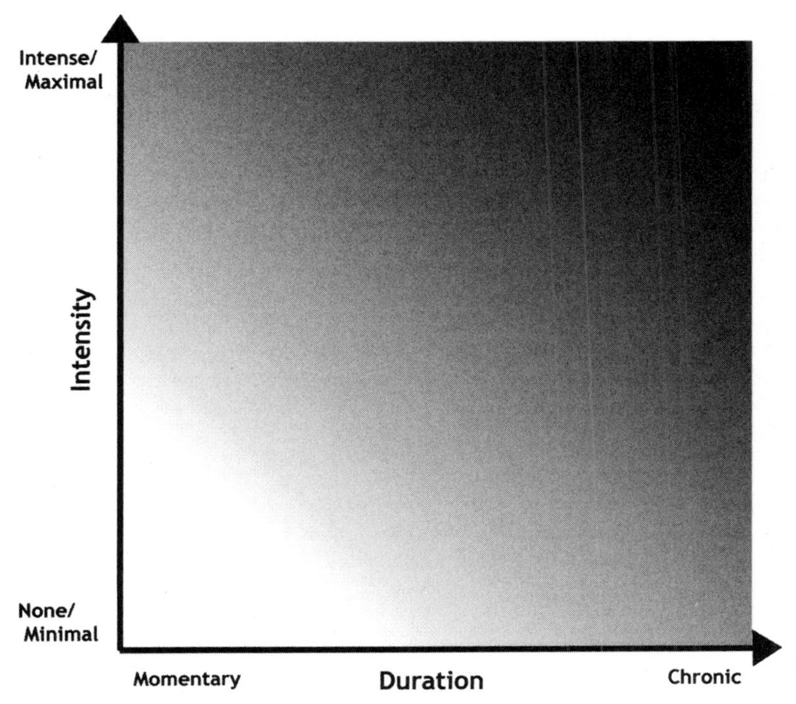

FIGURE 1-1 The two key aspects of pain relevant to refinement. The aversiveness of pain (darker shading = greater aversiveness) is primarily determined by duration and intensity: momentary and/or slight pain is less aversive than chronic and/or intense pain. Duration and intensity interact to affect aversiveness, although not in a simple additive way (the shading on this diagram does not imply a linear relationship). In humans, the aversiveness of pain is also affected by psychological factors, such as how controllable or predictable the pain is, and its context or consequences. There is little information about the influence of such effects in other animals (but see Chapter 4); thus for most practical purposes, the alleviation of pain in research animals typically means reducing its duration and/or its intensity, and both are refinements to be made whenever possible (see Chapters 3, 4, and 5).

cal concern and serious consideration of its alleviation. Chapters 3 and 4 address this in more detail.

An "unpleasant sensory and emotional experience" is at the core of the IASP definition of pain. Because sensory experiences and emotions (see Box 1-2) involve inner, private states that cannot be accessed directly by others, "[pain] is always subjective" (IASP 1979). This has some important practical implications. First, pain can never be measured directly, even when treating or researching human pain. Instead, the subjects' reports of

BOX 1-2
Emotion, Affect, Consciousness, and Awareness

In everyday use, "emotion" means a feeling that is consciously experienced and either negative (e.g., fear) or positive (e.g., joy). To scientists specializing in emotion research, states that are positive (i.e., accepted/preferred) or negative (i.e., aversive/not tolerated/avoided) are said to have a property called "valence." In the context of animal pain, the term "**affect**" instead of "emotion" is used because it is the scientific word whose meaning is closest to the colloquial use of "emotion," while also being less anthropomorphic. Thus to scientists specializing in emotion research, "affect" (or "affective") covers all states with valence; these include emotions (typically regarded as specific states induced by ongoing stimuli or events), moods (more generalized and longer lasting), and certain clinical conditions (e.g., depression) (Panksepp 2005; Rolls 2000, 2005; Russell 2003; Winkielman et al. 2005). However, some researchers use the terms specifically to refer to the human experience of conscious feelings (Panksepp 2005; Russell 2003).

This use of the terms "affect" and "affective" requires clarification of the terms "conscious" and "consciousness." The word "conscious" has a range of meanings, from the experience of the most basic form of sensation to the ability to have higher-order thoughts about one's own experiences, perspectives, or states of knowledge. In this report, **conscious** is used only to mean the former, thus referring to the "raw feel" of stimuli or events (Block 1997) or "the experience of sensation" (Merker 2007). Terms for this in specialized literatures include "qualia" (the inner "what it is like" aspects of, for example, seeing the color green or feeling angry; Tye 2007); "primary consciousness" (Edelman 2004); "qualitative consciousness" (van Gulick 2008); and "phenomenal consciousness" (Block 1997; Tye 2007). This basic form of consciousness is generally thought to be widely distributed across the animal kingdom (though how widely is a matter of debate; see text in this chapter). For brevity, in this report we use "consciousness" or "awareness" interchangeably. In the context of pain, such **awareness** is what distinguishes pain from nociception (see Boxes 1-3 and 1-4).

their own pain (e.g., via verbal descriptions or Likert scale values) are used as proxy measures (see Chapter 3). Such a report is the closest we have to a "gold standard." Second, in nonverbal organisms, be they laboratory animals or nonverbal humans such as babies, this type of self-report is not possible. As a result, making inferences about their pain is more challenging. Box 1-3 defines some key terms central to understanding these complex and essential aspects of pain, and the following section discusses further key challenges in understanding and identifying animal pain. Box 1-4 outlines approaches that come closest to these "gold standards" in animal research: that is, the closest one can come experimentally to self-report in nonverbal subjects.

BOX 1-3
Which (Unconscious) Nociceptive Responses May Not Indicate (Conscious) Pain?

Various models and examples can help identify responses to noxious stimuli that do not necessarily involve pain. Such responses occur (1) in organisms with either no nervous system or a nervous system so simple that scientists believe the organism is not capable of affect; (2) in mammals whose forebrains are not receiving input from the periphery; and (3) in humans whose pain has been suppressed (e.g., by analgesics/anesthetics).

Autonomic responses to noxious stimuli. In adult humans, postoperative cortisol output is undiminished by analgesics that successfully treat the reported pain (Schulze et al. 1988 cited by Lee et al. 2005; Dahl et al. 1992; see also Carrasco and Van de Kar 2003). Sympathetic responses such as tachycardia, hypertension, and pupil dilation occur in response to noxious stimuli in decerebrate rats and dogs (Sherrington 1906, reviewed in Sivarao et al. 2007).

Simple avoidance responses. Nonlearned avoidance responses are present in even simple single-celled organisms and require no affect (Rolls 2000; Tye 2007; Winkielman et al. 2005). The withdrawal of body parts (e.g., limbs, tails) from noxious stimuli also occurs in decerebrate cats (Sherrington 1906), and spinal-transected cats and rats in which connections to the brain are severed (e.g., Grau et al. 1998). In spinally transected cats, pinching or clamping the tail promotes stepping movements of the hindlimbs (Lovely et al. 1986), as though simple locomotory escape movements can occur even without pain. Some learned avoidance responses (e.g., classically conditioned withdrawal) have even been observed in the sea slug *Aplysia* (reviewed by Allen 2004). Other research reveals the instrumental learning of avoidance responses normally associated with pain with no possible involvement of the brain: spinally transected rats learn to keep their limbs withdrawn for longer periods of time if doing so will terminate the insult (Grau et al. 1998).

Other behavioral responses. Turning of the head and neck toward the noxious stimulus, some vocalization, and the licking of affected paws may occur in decerebrate animals (Baliki et al. 2005; King et al. 2003; Sherrington 1906).

Other responses. Cerebral blood flow increases during venipuncture in human fetuses as young as 16 weeks gestational age, even though the thalamocortical connections required for nociceptive input to reach the cortex have not developed (Lee et al. 2005). Isoflurane-anesthetized rats show activation in several forebrain regions (e.g., cingulate and insular cortices) in response to noxious stimuli applied to a paw (Hess et al. 2007).[a]

---

[a]The responses listed here are unreliable as indices of pain when attemtpting to assess whether a particular species or stage of development can experience pain and not just nociception. To make this assessment, stronger evidence is required. The absence of this stronger evidence is what fuels debates about nonmammalian vertebrates (see Box 1-4 and text). In intact animals, however, these nociceptive responses do often play an important and significant role in pain assessment (see Chapter 3).

BOX 1-4
Which Responses Indicate Pain and Which Nonhuman Vertebrates Display Them?

To determine whether animals can experience pain (not simply nociception), it is necessary to show that they can discriminate painful from nonpainful states; make decisions based on this discrimination in a way that cannot arise from evolved nonconscious nociceptive responses (cf. text and Box 1-3); demonstrate motivations to avoid pain; and display affective states of fear or anxiety if threatened with noxious stimuli. In addition, animals experiencing pain might be expected to exhibit spontaneous behavioral changes including sustained signals of distress and impairments in normal behaviors such as sleep (see text and Box 1-3).

The discrimination of painful states: evidence from operant experiments. In some learning paradigms, drug infusions are used as "discriminative stimuli," that is, experimental cues that predict which of two alternative learned operant responses will yield reward (e.g., whether a right or a left lever press will deliver food). In such experiments, rats show by shifting their operant response for food that they are able to distinguish injections of aspirin from injections of saline; furthermore, rats with arthritis learn this distinction more readily than do control rats (Weissman 1976; see also Colpaert 1978 and Swedberg et al. 1988). Thus, pain can serve as a discriminative stimulus, something the committee does not believe could occur without awareness.

Motivations to avoid pain or noxious stimuli. In learning paradigms in which an operant delivers an analgesic, rats in models-of-pain experiments lever press to self-medicate, and at a much higher rate than control animals. For example, rats with ligated spinal nerves lever press for clonidine, while controls do not (Martin et al. 2006). Rats, mice, primates, and pigeons also lever press to avoid electric shock (which may be painful depending on its intensity and duration; cf. Carlsson et al. 2006). Furthermore, oral self-administration of nonsteroidal anti-inflammatory drugs (NSAIDs) is observed in lame (i.e., arthritic) rats and chickens but not in their healthy counterparts (Colpaert et al. 1980; Danbury et al. 2000).

Similar research has not been conducted on reptiles, amphibians, or fish but frogs, tadpoles, and fish do show conditioned active avoidance responses when a cue is paired with shock (Dunlop et al. 2003; Overmier and Papini 1986; Strickler-Shaw and Taylor 1991). Fish display this response even if it involves swimming over a hurdle that

ANIMAL PAIN: DO ALL VERTEBRATES EXPERIENCE PAIN?

The general acceptance that many animal species can experience pain underlies the emphasis on pain in guidelines and laws on humane care (see Appendix B) as well as the scientific validity of using animals to investigate clinical pain (see Appendix A). However, the question of which species and/or developmental stages experience pain, and which instead merely display nociception (cf. Boxes 1-2 and 1-3), is a complex and sometimes controversial topic. Some observers argue that only humans, specifically offers resistance (Behrend and Bitterman 1962). Similarly, fish learn to avoid hooks in angling trials (Beukema 1970). However, it is not certain that such simple avoidance learning requires the experience of conscious pain (see text and Box 1-3).

Spontaneous behavioral changes. Noxious stimuli can cause vocalization (including ultrasonic calls in rodents) and signs of apparent apathy in mammals (see Chapter 3). Moreover, the use of inescapable electric shock to create mammal models of depression is well documented in the neuroscience literature. Sleep disruption (assessed via EEG activity) occurs in rats with arthritis or persistent neuropathic pain (Blackburn-Munro 2004). Although these responses seem inconsistent with mere nociception (see Box 1-3), it is not yet proven that they result from pain. For instance, while fish injected with acid or bee venom show suppressed feeding and other behavioral alterations (Ashley et al. 2009; Sneddon et al. 2003a,b), such changes are not universally accepted as indicative of pain (Rose 2002). Recent studies with fish have shown, however, that the brain is active during noxious stimulation (with the forebrain being the most significantly affected) and that this activity differs from that of nonnoxious stimuli (Dunlop and Laming 2005; Nordgreen et al. 2007; Reilly et al. 2008).

In summary, evidence for the conscious experience of pain is strong for mammals and birds, but conclusive studies are either in progress for other taxa such as fish or have not yet been conducted. Pending such needed research, this report treats all vertebrates as capable of experiencing pain (see text).[a]

---

[a]It is important to remember that there is scientific evidence to suggest that pain or the threat of noxious stimuli causes fear and/or anxiety. Much research shows that the mere threat of foot shock (i.e., the application of electric current on the foot) induces behavioral and physiological signs of stress in rats and mice that can be alleviated with compounds that reduce anxiety in humans (anxiolytics). Similar data are available for pigeons (Vanover et al. 2004). Furthermore, in one experiment an anxiety-inducing drug was used as a "discriminative stimulus" in pigs: the operant that would yield food was varied experimentally (e.g., from right lever to left lever) according to whether the subject was simultaneously infused with the anxiogenic drug or saline. Animals learned this discrimination successfully and performed a different operant for food depending on the compound of their infusion. The pigs were subsequently exposed to electric shock, which caused them to spontaneously select the anxiogenic rather than the saline operant when working for food. This finding suggests that the pigs' experience of the electric shock included the sensation of anxiety (Carey and Fry 1993). No such research has been conducted on reptiles, amphibians, or fish.

only humans past early infancy, experience pain (e.g., Carruthers 1996), while others suggest that all vertebrates, and some or even all invertebrates, are likely able to do so as well (Bateson 1991; Sherwin 2001; Tye 2007). Between these extremes lies a range of other, more generally accepted assessments.

With a focus on vertebrates, this section presents a brief discussion of what constitutes good evidence of the capacity to experience pain. The discussion emphasizes the strength of the evidence that all mammals (including rodents) are able to experience pain; raises the possibility that fish may feel pain; highlights the many things that are simply not known because the relevant research has not yet been conducted; and explains why the issue remains one of judgment rather than certainty. This section also lays the foundation for Chapter 3, on the recognition and assessment of pain.

There are two broad methods of assessing which animals can experience pain. The first is to demonstrate the presence of the anatomy and physiology that appear to be a requirement for pain in humans. The second is to investigate which species show responses to noxious stimuli suggestive of pain. Neither approach is adequate in itself, as noted below, but they are complementary and each informs the other.

The anatomy and physiology of human pain are well understood: the nature of nociceptive inputs and circuits is well characterized, and specific forebrain regions (e.g., the insular, prefrontal, and anterior cingulate cortices) have been implicated in the experience of pain (see Baliki et al. 2006 and Chapter 2). Several authors have used this knowledge to catalogue similarities and differences between humans and other species (Allen 2004, 2006; Bateson 1991; Rose 2002; Sneddon 2006; Varner 1999). They typically highlight homologies both in structure and in responses to noxious stimuli in the forebrains of humans and other mammals such as rats (see Apkarian et al. 2006; Borsook et al. 2006, 2007). Other vertebrates—birds, reptiles, fish, and amphibians—have peripheral and spinal nociceptive circuitry akin to that of humans, but not the specific forebrain regions involved in human pain. Invertebrates share still fewer similarities with humans—principally, only nociceptors and certain neurotransmitters (Allen 2004; Allen et al. 2005).

The challenge in interpreting such data is knowing what emphasis to place on the various elements. Which, if any, underlie pain? Even the argument that certain forebrain structures are required for pain (Rose 2002) is problematic because it presupposes a complete understanding of how and where pain is generated in the human brain, when in fact this is still under study (the anterior cingulate, for instance, is activated by subliminal stimuli—i.e., stimuli of which humans are unaware—as well as by pain; Kilgore and Yurgelun-Todd 2004; Sidhu et al. 2004; Box 1-3). Such an argument also assumes that, evolutionarily, any cortical subregions involved in pain became so only *after* their specialization into these subregions (thus ignoring the possible functions of these regions' evolutionary precursors). Furthermore, it does not clarify the states of animals whose nervous systems differ greatly from that of humans but may still have analogous structures and functions (e.g., invertebrates, which lack a central nervous system, and birds or fish, which have complex forebrains but no neocortex; Allen 2004; Shriver 2006). This type of uncertainty is one reason the phylogenetic distribution of pain is a matter of discussion and debate.

Despite these ongoing debates, it is generally agreed that, in mammals, pain does require a cortex (though see Merker 2007 for an opposing view). Therefore, it is typically assumed that any responses in, for example, decerebrate mammals cannot be used reliably to identify which species or developmental stages feel pain (Box 1-3). The second way to determine which animals experience pain is by examining their physiological and behavioral responses to noxious stimuli.

Pain in humans is associated with a range of physiological and behavioral responses. Some are best described as nociceptive because they occur in response to noxious stimuli even when pain is suppressed by analgesia or anesthesia (Box 1-3). But humans can also assess and report the presence or absence of pain, describe its qualities, and use this information to make decisions (e.g., when to seek help, when to take analgesics, or which pain management strategy to adopt). Pain also leads to the protection and "nursing" of affected regions. Such behaviors reflect a strong, sustained desire to minimize or end pain (it has been argued that the affective component of pain is essential for the way it strongly motivates escape and avoidance; van Gulick 2008; McMillan 2003).[2] As recent studies have demonstrated, postsurgical/postprocedural, persistent, or chronic pain can have deleterious effects on behavior, cognition, and brain function (e.g., problems with sleep, attention, or depression, even possible loss of gray matter; Apkarian et al. 2004a,b). These findings suggest several useful indices for identifying animals that experience pain, not simply nociception (Box 1-4). Unfortunately, data on these key variables for many animal species have not been collected, generally because the research is methodologically challenging (Box 1-4). This is another reason why the phylogenetic distribution of pain is a matter of discussion and debate.

Although definitive evidence is often unavailable, this report does not treat the absence of evidence as evidence of absence. Instead, the consensus of the committee is that all vertebrates should be considered capable of experiencing pain. This judgment is based on the following two premises: (1) the strong likelihood that this is correct, particularly for mammals and birds (Box 1-4 provides compelling evidence for rats, for example); and (2) the consequences of being wrong, that is, acting on the assumption that all vertebrates are not able to experience pain and so treating pain as though it were merely nociception, an error with obvious and serious ethical implications. This report, therefore, considers nociceptive responses in vertebrates as likely indices of pain rather than nonconscious responses to noxious stimuli.

---

[2]As explained in Chapter 4, the protective role of pain is one reason that complete elimination of postoperative pain may not be desirable.

CAUSES OF PAIN IN RESEARCH ANIMALS

Understanding the potential causes of pain in research animals can facilitate the anticipation or recognition of both the types of specific stimuli or tissue responses and the situations (in terms of management, husbandry, or experiment) in which pain is likely.

As a general guideline to types of stimuli or tissue responses that cause pain in animals, many codes and recommendations state something like the following: "Unless the contrary is established, investigators should consider that procedures that cause pain or distress in human beings may cause pain or distress in other animals" (Principle #4, US Government Principles for the Utilization and Care of Vertebrate Animals Used in Testing, Research, and Teaching; IRAC 1985); or "[a painful procedure is] any procedure that would reasonably be expected to cause more than slight or momentary pain and/or distress in a human being to which the procedure is applied" (USDA Policy #11; see Kohn et al. 2007 for a similar view from the American College of Laboratory Animal Medicine).

The committee agrees with these statements, but cautions that in humans the type and intensity of stimuli detected by nociceptors differ for different tissues (as outlined previously in this chapter). For example, cutting, crushing, or burning skin reliably causes pain, whereas these same stimuli applied to the wall of a hollow organ rarely cause pain (see Ness and Gebhart 1990 for a review). If this is true for a single species, it is not hard to imagine the differences that may exist across the tissues of different species, especially those that have evolved to live in very different worlds (e.g., very hot or cold environments) or to have very different sensory abilities (e.g., abilities to detect ultrasound or electromagnetic fields; Allen 2004). Indeed, species-specific differences in response to painful events are well documented (Paul-Murphy et al. 2004; Valverde and Gunkel 2005). There is also variation in response to drugs that are analgesic in one species but not in another; for example, the effects of opioids are very unpredictable in birds (Hughes and Sufka 1991). For all these reasons, one *cannot assume* that what causes pain in humans will do so in all other organisms, and conversely, that what does not cause humans pain is equally benign in all other organisms. Thus it is essential to assess pain in an animal on a case-by-case basis (see Chapter 3).

Examples of stimuli or tissue injury that cause pain in research animals, whether from disease conditions or experimental procedures, are given in Table 1-1. They are broadly broken down by tissue type, to mirror the tissue-specific noxious stimuli listed in the section above on nociception. The list in Table 1-1 is intended to be illustrative, not all-inclusive. Note that when assessed using the techniques discussed in Chapter 3, the aversiveness of the pain resulting from each item in the table can vary greatly (typically

TABLE 1-1 Examples of Painful Procedures or Conditions by Type and Anatomic Location

Abdominal	Peritonitis, pancreatitis, hepatitis, cholelithiasis, distension of viscera, bowel obstruction, visceral tumors, laparotomy
Cardiothoracic	Myocarditis, pneumonitis, myocardial infarction, pneumonia, bronchitis, vasculitis, vascular grafts, thoracotomy
Dermatologic	Pruritis, chemical and thermal burns, cellulitis, otitis, skin tumors, incision, needle puncture
Musculoskeletal	Restraint, arthritis, periostitis, ischemia, application of a tourniquet, tendonitis, inflammation of joints, deep chemical or thermal burns, crush, bruising, necrosis, fracture, bone graft harvest, bone tumor, osteotomy, incision, craniectomy, degenerative joint disease
Neurologic	Encephalitis, meningitis; crush, ligation, or transection of nerves; tumor of neural tissue; neuroma
Ocular	Glaucoma, uveitis, corneal ulcer, orbital blood sampling, ocular tumor
Orofacial	Oral tumors, temporomandibular joint disease, gingivitis, tooth extraction, pulpotomy, tooth abscess
Systemic	Sepsis, sickness syndrome, autoimmune diseases
Urogenital	Pyelonephritis, cystitis, acute renal failure, ureteral or urethral obstruction, pyometra, urinary catheterization, mastitis, ovariohysterectomy, castration, urogenital tumor, dystocia

from mild to severe), depending on its duration and intensity. Again, case-by-case assessment and treatment are critical and essential (see Chapters 3 and 4).

In the context of animals used in research and testing, the following circumstances will or are likely to cause pain[3]:

Non-research-related disease or injury: Tissue damage and/or inflammation (e.g., injuries sustained in fighting with conspecifics, ammonia burns from soiled litter), mastitis, abscesses and other infections, arthritis and other diseases resulting from aging, and parturition.

Husbandry or veterinary treatment: Invasive procedures as part of normal husbandry, preparation for research, or before the animal's designation as a research subject (e.g., castration, dehorning, teeth clipping, tail docking, tail-tip removal for genotyping, ear notching, microchip implantation, catheter placement, injection).

---

[3]It is important to remember that early postnatal tissue injury can alter adult nociceptive processing, including enhanced responses to noxious stimuli.

Research byproduct: Research on disease (infectious or noninfectious, such as cancer), toxins, tissue damage (e.g., burns, bone breakage), some aspects of drug dependence (e.g., opiate withdrawal that causes lower back and/or abdominal pain and cramps); and surgery, in which pain may be a consequence of research but is neither an element of the research nor a focus of study. Hyperalgesia may also occur as a result of "sickness syndrome" (see Chapter 4).

The use of pain as a tool to motivate or shape behavior: Noxious stimuli (e.g., foot shock) for the purposes of training or motivation during behavioral experiments (punishment/negative reinforcement), for the experimental assessment of fear (e.g., in fear-conditioning paradigms), or for the experimental induction of depression-like states.

Pain as the focus of research: For a review and description of common animal models of persistent pain, including humane endpoints for this type of research, see Appendix A.

These five circumstances may involve pain that differs in causation, duration, and intensity. They also vary in the nature and defensibility of the justification for inducing that pain and for allowing it to be untreated, as discussed below.

IS PAIN IN ANIMALS EVER JUSTIFIABLE?

According to current US laws and guidelines, some animal pain is justified in some circumstances. For example, USDA Policy #12 states that "a description of procedures or methods designed to assure that discomfort and pain to animals will be limited to that which is unavoidable in the conduct of scientifically valuable research" (USDA 1997b), the Public Health Service Policy on Humane Care and Use of Laboratory Animals (DHHS 2002) mandates that "procedures which may cause more than momentary or slight pain or distress to animals should be performed with appropriate sedation, analgesia, or anesthesia, unless the procedure is justified for scientific reasons in writing by the investigator," and section 2.31(e) of the US Animal Welfare Act states that "A description of procedures designed to assure that discomfort and pain to animals will be limited to that which is unavoidable for the conduct of scientifically valuable research" (AWA 1990).

Thus there exist situations in which pain and/or the withholding of analgesic drugs can be justified scientifically. As noted above, such situations include the use of noxious stimuli as a tool to motivate or shape behavior or the study of pain as the focus of research (see Appendix A). However, as indicated at the beginning of this chapter, the ethical justification for such research should consider both the costs to the animal and the expected benefits of the research to humankind (although a small research component may directly benefit the animals themselves, for example, in the development of better analgesics for rats or mice; for an in-depth ethical analysis, see "Animal Welfare Considerations of Research with Persistent Pain Models" in Chapter 4). Consistent with the concerns of the general public (Kohn et al. 2007) it is the view of this committee that the greater the cost to the research animals in terms of pain, distress, and negative impact on welfare and well-being, the stronger the scientific justification of the research should be.

CONCLUSIONS AND RECOMMENDATIONS

1. Although there is general agreement that pain is an aversive state experienced by mammals and probably all vertebrates, the committee assumes in this report that *all* vertebrates are capable of experiencing pain.
2. The assumption of similarities in pain between humans and animals is a useful rule of thumb. However, the scientific outcomes should be taken into account when the 4th Government Principle is interpreted.
3. Pain in research animals may be induced deliberately as part of a research procedure (e.g., when pain is the subject of research) or may be an unintended byproduct of other research objectives, husbandry, or other factors.
4. As was emphasized in the Distress report (NRC 2008), the *Three Rs* (replacement, refinement, and reduction) should be the standard for identifying, modifying, avoiding, and minimizing most causes of pain in laboratory animals. While research on pain or on methods of alleviating pain may unavoidably cause animal distress and severe perturbation of animal welfare, the goal of researchers, veterinary teams, and IACUCs should be to reduce and alleviate pain in laboratory animals to the minimum necessary to achieve the scientific objective.

REFERENCES

Allen C. 2004. Animal pain. Noûs 38:617-643.

Allen C. 2006. Animal consciousness. The Stanford Encyclopedia of Philosophy (Winter Edition). Available at http://plato.stanford.edu/archives/win2006/entries/consciousness-animal/. Accessed June 3, 2008.

Allen C, Fuchs PN, Shriver A, Wilson HD. 2005. Deciphering animal pain. In: Ayded M, ed. Pain: New Essays on Its Nature and the Methodology of Its Study. Cambridge, MA: MIT Press.

Apkarian AV, Sosa Y, Krauss BR, Thomas PS, Fredrickson BE, Levy RE, Harden RN, Chialvo DR. 2004a. Chronic pain patients are impaired on an emotional decision-making task. Pain 108(1-2):129-136.

Apkarian AV, Sosa Y, Sonty S, Levy RM, Harden RN, Parrish TB, Gitelman DR. 2004b. Chronic back pain is associated with decreased prefrontal and thalamic gray matter density. J Neurosci 24(46):10410-10415.

Apkarian AV, Lavarello S, Randolf A, Berra HH, Chialvo DR, Besedovsky HO, del Rey A. 2006. Expression of IL-1beta in supraspinal brain regions in rats with neuropathic pain. Neurosci Lett 407(2):176-181.

Ashley PJ, Ringrose S, Edwards KL, Wallington E, McCrohan CR, Sneddon LU. 2009. Effect of noxious stimulation upon anti-predator responses and dominance status in rainbow trout. Anim Behav 77(2):403-410.

AWA (Animal Welfare Act). 1990. Animal Welfare Act. Available at www.nal.usda.gov/awic/legislat/awa.htm. Accessed June 9, 2008.

Baliki M, Calvo O, Chialvo DR, Apkarian AV. 2005. Spared nerve injury rats exhibit thermal hyperalgesia on an automated operant dynamic thermal escape task. Mol Pain 1:18.

Baliki MN, Chialvo DR, Geha PY, Levy RM, Harden RN, Parrish TB, Apkarian AV. 2006. Chronic pain and the emotional brain: Specific brain activity associated with spontaneous fluctuations of intensity of chronic back pain. J Neurosci 26(47):12165-12173.

Bateson P. 1991. Assessment of pain in animals. Anim Behav 42:827-839.

Behrend ER, Bitterman ME. 1962. Avoidance-conditioning in the goldfish: Exploratory studies of the CS-US interval. Am J Psychol 75:18-34.

Beukema JJ. 1970. Acquired hook-avoidance in the pike *Esox lucius L.* fished with artificial and natural baits. J Fish Biol 2:155-160.

Blackburn-Munro G. 2004. Pain-like behaviours in animals: How human are they? Trends Pharmacol Sci 25(6):299-305.

Block NJ. 1997. Begging the question against phenomenal consciousness. In: Block NJ, Flanagan OJ, Güzeldere G, eds. The Nature of Consciousness. Cambridge, MA: MIT Press. p 175-179.

Borsook D, Becerra L, Hargreaves R. 2006. A role for fMRI in optimizing CNS drug development. Nat Rev Drug Discov 5(5):411-424.

Borsook D, Pendse G, Aiello-Lammens M, Glicksman M, Gostic J, Sherman S, Korn J, Shaw M, Stewart K, Gostic R, Bazes S, Hargreaves R, Becerra L. 2007. CNS response to a thermal stressor in human volunteers and rats may predict the clinical utility of analgesics. Drug Develop Res 68(1):23-41.

Carey MP, Fry JP. 1993. A behavioural and pharmacological evaluation of the discriminative stimulus induced by pentylenetetrazole in the pig. Psychopharmacology (Berl) 111(2):244-250.

Carlsson K, Andersson J, Petrovic P, Petersson KM, Ohman A, Ingvar M. 2006. Predictability modulates the affective and sensory-discriminative neural processing of pain. Neuroimage 32(4):1804-1814.

Carrasco GA, Van de Kar LD. 2003. Neuroendocrine pharmacology of stress. Eur J Pharmacol 463(1-3):235-272.

Carruthers P. 1996. Language, Thought and Consciousness. Cambridge: Cambridge University Press.

Colpaert FC. 1978. Discriminative stimulus properties of narcotic analgesic drugs. Pharmacol Biochem Behav 9(6):863-887.

Colpaert FC, De Witte P, Maroli AN, Awouters F, Niemegeers CJ, Janssen PA. 1980. Self-administration of the analgesic suprofen in arthritic rats: Evidence of *Mycobacterium butyricum*-induced arthritis as an experimental model of chronic pain. Life Sci 27(11): 921-928.

Dahl JB, Rosenberg J, Kehlet H. 1992. Effect of thoracic epidural etidocaine 1.5% on somatosensory evoked potentials, cortisol and glucose during cholecystectomy. Acta Anaesthesiol Scand 36(4):378-382.

Danbury TC, Weeks CA, Chambers JP, Waterman-Pearson AE, Kestin SC. 2000. Self-selection of the analgesic drug carprofen by lame broiler chickens. Vet Rec 146(11):307-311.

DHHS (Department of Health and Human Services). 2002. Public Health Service Policy on Humane Care and Use of Laboratory Animals. Available http://grants.nih.gov/grants/olaw/references/phspol.htm. Accessed June 9, 2008.

Dunlop R, Laming P. 2005. Mechanoreceptive and nociceptive responses in the central nervous system of goldfish (*Carassius auratus*) and trout (*Oncorhynchus mykiss*). J Pain 6:561-568.

Dunlop R, Millsop S, Laming P. 2003. Avoidance learning in goldfish (*Carassius auratus*) and trout (*Oncorhynchus mykiss*) and implications for pain perception. Appl Anim Behav Sci 97:255-271.

Edelman GM. 2004. Wider Than the Sky: The Phenomenal Gift of Consciousness. New Haven: Yale University Press.

Gentle MJ. 2001. Attentional shifts alter pain perception in the chicken. Anim Welf 10(Suppl 1):187-194.

Grau JW, Barstow DG, Joynes RL. 1998. Instrumental learning within the spinal cord: I. Behavioral properties. Behav Neurosci 112(6):1366-1386.

Hess A, Sergejeva M, Budinsky L, Zeilhofer HU, Brune K. 2007. Imaging of hyperalgesia in rats by functional MRI. Eur J Pain 11(1):109-119.

Hughes RA, Sufka KJ. 1991. Morphine hyperalgesic effects on the formalin test in domestic fowl (*Gallus gallus*). Pharmacol Biochem Behav 38(2):247-251.

IASP (International Association for the Study of Pain). 1979. IASP Pain Terminology. Available at www.iasp-pain.org/AM/Template.cfm?Section=Pain_Definitions&Template=/CM/HTMLDisplay.cfm&ContentID=1728#Pain. Accessed January 8, 2009.

IRAC (Interagency Research Animal Committee). 1985. The U.S. Government Principles for the Utilization and Care of Vertebrate Animals Used in Testing, Research, and Training. Federal Register Vol. 50, No. 97 (May 20, 1985). Available at http://grants.nih.gov/grants/olaw/references/phspol.htm#USGovPrinciples. Accessed June 9, 2008.

Keogh E, Cochrane M. 2002. Anxiety sensitivity, cognitive biases, and the experience of pain. J Pain 3(4):320-329.

Kilgore WD, Yurgelun-Todd DA. 2004. Activation of the amygdala and anterior cingulate during nonconscious processing of sad versus happy faces. Neuroimage 21(4):1215-1223.

King CD, Devine DP, Vierck CJ, Rodgers J, Yezierski RP. 2003. Differential effects of stress on escape and reflex responses to nociceptive thermal stimuli in the rat. Brain Res 987(2):214-222.

Kohn DF, Martin TE, Foley PL, Morris TH, Swindle MM, Vogler GA, Wixson SK. 2007. Public statement: Guidelines for the assessment and management of pain in rodents and rabbits. J Am Assoc Lab Anim Sci 46(2):97-108.

Langford DJ, Crager SE, Shehzad Z, Smith SB, Sotocinal SG, Levenstadt JS, Chanda ML, Levitin DJ, Mogil JS. 2006. Social modulation of pain as evidence for empathy in mice. Science 312(5782):1967-1970.

Lee SJ, Ralston HJ, Drey EA, Partridge JC, Rosen MA. 2005. Fetal pain: A systematic multidisciplinary review of the evidence. JAMA 294(8):947-954.

Lovely RG, Gregor RJ, Roy RR, Edgerton VR. 1986. Effects of training on the recovery of full-weight-bearing stepping in the adult spinal cat. Exp Neurol 92(2):421-435.

Martin TJ, Kim SA, Eisenach JC. 2006. Clonidine maintains intrathecal self-administration in rats following spinal nerve ligation. Pain 125(3):257-263.

McMillan FD. 2003. A world of hurts: Is pain special? JAVMA 223(2):193-196.

Merker B. 2007. Consciousness without a cerebral cortex: A challenge for neuroscience and medicine. Behav Brain Sci 30(1):63-81; discussion 81-134.

Ness TJ, Gebhart GF. 1990. Visceral pain: A review of experimental studies. Pain 41(2): 167-234.

Nordgreen J, Horsberg TE, Ranheim B, Chen C. 2007. Somatosensory evoked potentials in the telencephalon of Atlantic salmon (*Salmo salar*) following galvanic stimulation of the tail. J Comp Physiol A 193:1235-1242.

Overmier JB, Papini MR. 1986. Factors modulating the effects of teleost telencephalon ablation on retention, relearning, and extinction of instrumental avoidance behavior. Behav Neurosci 100(2):190-199.

Panksepp J. 2005. Affective consciousness: Core emotional feelings in animals and humans. Conscious Cogn 14(1):30-80.

Paul-Murphy J, Ludders JW, Robertson SA, Gaynor JS, Hellyer PW, Wong PL. 2004. The need for a cross-species approach to the study of pain in animals. JAVMA 224(5):692-697.

Price DD. 2002. Central neural mechanisms that interrelate sensory and affective dimensions of pain. Mol Interv 2(6):392-403, 339.

Reilly SC, Quinn JP, Cossins AR, Sneddon LU. 2008. Novel candidate genes identified in the brain during nociception in common carp. Neuro Sci Letts 437:135-138.

Rolls ET. 2000. Precis of the brain and emotion. Behav Brain Sci 23(2):177-191; discussion 192-233.

Rolls ET. 2005. Emotion Explained. Oxford, New York: Oxford University Press.

Rose JD. 2002. The neurobehavioral nature of fishes and the question of awareness and pain. Rev Fish Sci 10:1-38.

Russell JA. 2003. Core affect and the psychological construction of emotion. Psychol Rev 110(1):145-172.

Russell WMS, Burch RL. 1959. The Principles of Humane Experimental Technique. London: Methuen.

Sherrington CS. 1906. The Integrative Action of the Nervous System. New York: Charles Scribner's Sons.

Sherwin CM. 2001. Can invertebrates suffer? Or, how robust is argument-by-analogy? Animal Welfare 10:103-118.

Shriver A. 2006. Minding mammals. Philos Psychol 19(4):433-442.

Sidhu H, Kern M, Shaker R. 2004. Absence of increasing cortical fMRI activity volume in response to increasing visceral stimulation in IBS patients. Am J Physiol Gastrointest Liver Physiol 287(2):G425-G435.

Sivarao DV, Langdon S, Bernard C, Lodge N. 2007. Colorectal distension-induced pseudoaffective changes as indices of nociception in the anesthetized female rat: Morphine and strain effects on visceral sensitivity. J Pharmacol Toxicol Methods 56(1):43-50.

Sneddon LU. 2006. Ethics and welfare: Pain perception in fish. B Eur Assoc Fish Pat 26(1): 6-10.

Sneddon LU, Braithwaite VA, Gentle MJ. 2003a. Do fish have nociceptors: Evidence for the evolution of a vertebrate sensory system. Proc R Soc Lond B Biol Sci 270:1115-1122.

Sneddon LU, Braithwaite VA, Gentle MJ. 2003b. Novel object test: Examining pain and fear in the rainbow trout. J Pain 4:431-440.

Strickler-Shaw S, Taylor DH. 1991. Lead inhibits acquisition and retention learning in bullfrog tadpoles. Neurotoxicol Teratol 13(2):167-173.

Swedberg MD, Shannon HE, Nickel B, Goldberg SR. 1988. Pharmacological mechanisms of action of flupirtine: A novel, centrally acting, nonopioid analgesic evaluated by its discriminative effects in the rat. J Pharmacol Exp Ther 246(3):1067-1074.

Tye M. 2007. Qualia. The Stanford Encyclopedia of Philosophy (Fall Edition). Available at http://plato.stanford.edu/archives/fall2007/entries/qualia/. Accessed June 9, 2008.

USDA (United States Department of Agriculture). 1997a. APHIS Policy #11, Painful Procedures (issue dated: April 14, 1997). Available at: www.aphis.usda.gov/animal_welfare/downloads/policy/policy11.pdf. Accessed June 9, 2008.

USDA. 1997b. APHIS Policy #12, Considerations of Alternatives to Painful/Distressful Procedures (issue dated: June 21, 2000). Available at www.aphis.usda.gov/animal_welfare/downloads/policy/policy12.pdf. Accessed June 9, 2008.

Valverde A, Gunkel CI. 2005. Pain management in horses and farm animals. J Vet Emerg Crit Car 15(4):295-307.

Van Gulick R. 2008. Consciousness. The Stanford Encyclopedia of Philosophy (Spring Edition). Available at http://plato.stanford.edu/archives/spr2008/entries/consciousness/. Accessed June 9, 2008.

Vanover KE, Zhang L, Barrett JE. 2004. Discriminative stimulus and anxiolytic-like effects of the novel compound CL 273,547. Exp Clin Psychopharmacol 2(3):223-233.

Varner G. 1999. How facts matter: On the language condition and the scope of pain in the animal kingdom. Pain Forum 8(2):84-86.

Weissman A. 1976. The discriminability of aspirin in arthritic and nonarthritic rats. Pharmacol Biochem Behav 5(5):583-586.

Winkielman P, Berridge KC, Wilbarger JL. 2005. Emotion, behavior and conscious experience: Once more without feeling. In: Barrett LF, Niedenthal PM, Winkielman P, eds. Emotion and Consciousness. New York: Guilford Press.

2

Mechanisms of Pain

This chapter provides an analysis of the differences between nociception and pain, on the basis of the anatomy of the peripheral and central nervous systems and the role of nociceptors in pain perception. It includes discussion of the concept of persistent pain and presents information on the embryologic origins of pain. Finally it addresses the modulatory role of anxiety, fear, and stress on pain.

NOCICEPTION OR PAIN

Before discussing the anatomical and physiological bases for the generation of pain, it is important to reiterate the difference between nociception and pain. Nociception refers to the peripheral and central nervous system (CNS) processing of information about the internal or external environment, as generated by the activation of nociceptors. Typically, noxious stimuli, including tissue injury, activate nociceptors that are present in peripheral structures and that transmit information to the spinal cord dorsal horn or its trigeminal homologue, the nucleus caudalis. From there, the information continues to the brainstem and ultimately the cerebral cortex, where the perception of pain is generated (Figure 2-1).

Pain is a product of higher brain center processing, whereas nociception can occur in the absence of pain. For example, the spinal cord of an individual who suffered a complete spinal cord transection can still process information transmitted by nociceptors, but because the information cannot be transmitted beyond the transection stimulus-evoked pain is unlikely (see Chapter 1 for additional discussion).

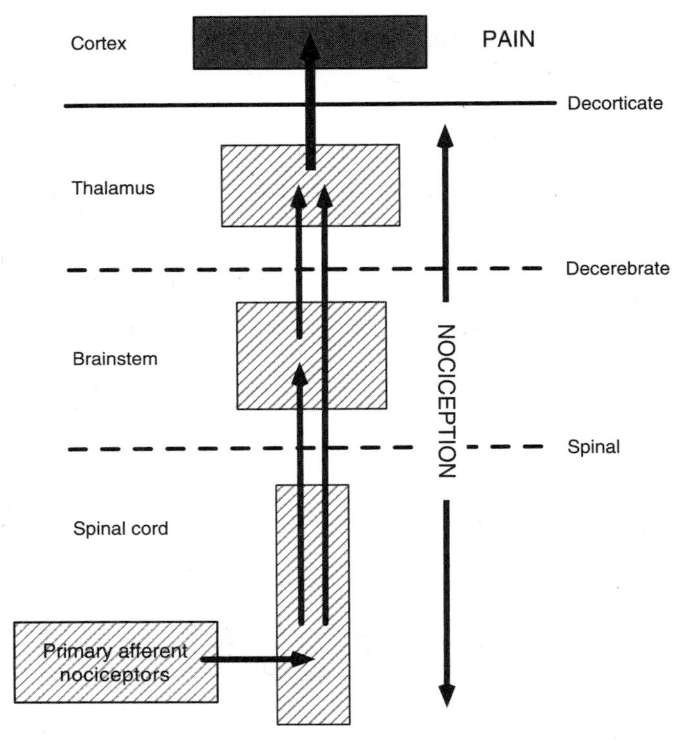

FIGURE 2-1 Anatomical distribution of nociception and pain. This figure schematizes the major neuroanatomical structures that differentiate nociception and pain, an understanding of which is essential for studies in which the animals may experience pain. Nociception refers to the process through which information about peripheral stimuli is transmitted by primary afferent nociceptors to the spinal cord, brainstem, thalamus, and subcortical structures. In contrast, the experience of pain can result only when there is activity of thalamocortical networks (represented in the dark shaded box at the top) that process the information conveyed by pathways of nociception. The magnitude of pain is determined to a great extent by the strength of descending inhibitory and facilitatory controls (in the lighter shaded boxes) that originate throughout the neuraxis and regulate the processing of ascending nociceptive messages. The figure also illustrates several important surgical preparations used to study nociceptive processing under conditions in which different parts of the brain are disconnected from afferent nociceptive input. Thus, transection of the spinal cord produces a "spinal" preparation. Decerebrate preparation entails transection of the brain between the midbrain (at the level of the colliculi) and the thalamus. In the decorticate preparation, connections from the thalamus to the cortex are severed. In all of these conditions, information generated by the activity of nociceptors located below the level of transection is unlikely to reach structures above the transection. No evidence exists at present that hormonal or other nonneural mechanisms are able to "bypass" the transection to access the brain and evoke a pain perception.

The distinction between nociception and pain is also important for behavioral studies in which an understanding of pain mechanisms is the ultimate goal. Many behavioral tests involve assessment of reflex responses to noxious stimuli, typically applied at threshold or just suprathreshold intensities (such as heating of the tail or the hindpaw) to incite a brief withdrawal of the tail (e.g., in the tail flick test) or paw. These are principally tests of nociceptive processing because stimulus duration is limited by the animal's response (e.g., a nociceptive withdrawal reflex). On the other hand, the endpoints of more complex behaviors (e.g., those involved in operant tests) are presumed to involve supraspinal areas of the brain and as such are tests of both nociception and pain. In that respect, operant tests in which animals perform a particular behavior (e.g., press a bar) to escape a stimulus provide information about both nociceptive processing and pain (see also Box 1-4 in Chapter 1).

Mechanisms of Nociception and Pain

Nociceptors

The anatomical basis for the generation of momentary pain is very well understood (Basbaum and Jessell 2000). Nociceptors are unusual neurons because they have a cell body with a peripheral axon and terminal (ending) that responds to the stimulus and a central branch that carries the information into the CNS. Briefly, there are two major classes of nociceptors that respond to different modalities of noxious stimuli.

The largest group of nociceptors is associated with unmyelinated axons, also called C-fibers, that conduct slowly and that respond to noxious thermal, mechanical, or chemical stimulation. Proteins in the membrane of these nociceptors transduce natural thermal, mechanical, or chemical stimulus energy into electrical impulses, which in turn are propagated along the peripheral and central axon of the nociceptor into the CNS (the spinal cord for the body and the trigeminal nucleus for the head). Importantly, biochemical and molecular analysis of the nociceptor has identified many of the transducer molecules that are activated by noxious stimuli, such as TRPV1, which responds to noxious heat, reduced pH as occurs in inflammation, and the chemical capsaicin. Another channel, TRPM8, responds to cold (Julius and Basbaum 2001). Many of these molecules are targets for therapeutic intervention in clinical pain conditions.

The second major nociceptor population is associated with thinly myelinated axons (A-delta fibers). These nociceptors conduct more rapidly than do unmyelinated C-fibers and likely convey "fast" (or sharp) momentary pain, as opposed to slow, diffuse pain, which is transmitted by the C-fibers.[1]

There is yet one more category of nociceptors characterized by unique properties. "Sleeping" or "silent" nociceptors are typically unresponsive to noxious intensities of mechanical stimulation except at extreme ranges of intensity. Although silent nociceptors are difficult to activate within the normal range of noxious stimulus intensities, after tissue insult these nociceptors "wake up" in response to endogenous chemical mediators associated with tissue injury. Silent nociceptors are typically associated with increased spontaneous activity and responsiveness to noxious and even innocuous stimulus intensities.

Spontaneous activity in nociceptors (whether A-delta, C-, or silent) is undesirable and pain producing; moreover, awakening silent nociceptors creates essentially new, additional nociceptive input to the CNS. All nociceptors have the capacity to sensitize. When they become more easily excitable (i.e., the threshold for activation is lowered), hyperalgesia (an increased response to a noxious stimulus) with or without allodynia develops and normally innocuous stimuli may provoke pain, thus directly affecting animal welfare. The consequences of such activities are discussed below in the section on persistent pain.

The Central Nervous System

The central branch of the nociceptor terminates in the dorsal horn of the spinal cord (or its trigeminal homologue in the brainstem), where it makes synaptic connections with a complex array of neurons that play different roles in nociceptive processing and pain. Some interneurons make connections with motor neurons that generate nociceptive withdrawal reflexes. Output neurons of the spinal cord, on the other hand, project rostrally and transmit the nociceptive message to the brainstem reticular formation and thalamus. Among the ascending pathways arising from the spinal cord (and its trigeminal homologue) are the spinothalamic and spinoreticulothalamic tracts, as well as the spinoparabrachial-amygdala pathway, which provides more direct access to limbic emotional circuits in the brain (via the amygdala) (Basbaum and Jessell 2000). Note that there is not a unitary pathway for generation of the affective component of the pain experience. Rather it is likely that different aspects of the nociceptive message are conveyed via different pathways and widely distributed to the cerebral cortex from the reticular formation, thalamus, and amygdala.

[1]Whereas virtually all nociceptors are A-delta and C-fibers, not all A-delta and C-fibers are nociceptors. It is thus both inaccurate and incorrect to generically refer to C-fibers as "pain" fibers.

Until recently, remarkably little was understood about the cortical mechanisms that underlie the perception of pain. Although electrophysiological studies have demonstrated that some neurons in the cortex respond to noxious stimuli, the extent to which this response represents or even correlates with pain was not clear. The development of powerful imaging methods, however, has provided critical information about the cortical processing of pain-related information (e.g., Apkarian et al. 2005; Bingel and Tracey 2008; Tracey and Mantyh 2007) and revealed that pain is not processed in a single area of the brain. Rather, the activity of different regions of the cortex underlies various features of the pain percept and cognitive recall for responses or emotional reactions. This information comes largely from human studies, in which a verbal correlate of pain perception is possible. For example, activity in the somatosensory cortices (S1 and S2) correlates best with the sensory-discriminative properties of the stimulus (e.g., location and intensity), and the affective components of the pain experience correlate with activity in the anterior cingulate gyrus and the insular cortex. Unfortunately, the activity of these regions cannot be used as a biomarker for pain, as it can also be generated by conditions that are clearly not painful (for additional discussion see Chapter 1).

Further Comments on the Distinction between Nociception and Pain

An unusual model to investigate the brain circuitry involved in nociception and pain was developed at the beginning of the 20th century by Charles Sherrington (1906), who appreciated early on the distinction between nociception and pain. Use of a "decerebrate preparation" (*cerveau isolé*) in laboratory animal research was more common years ago, but it remains useful for recording the activity of spinal cord or brainstem neurons under conditions not compromised by anesthetics or analgesics. With the animals under deep general anesthesia, the procedure involves transection of the brainstem at the level of the midbrain (typically between the inferior and superior colliculi), after which the rostral part of the brain (particularly subcortical structures and the cortex) no longer receives direct neuronal input from the spinal cord or brainstem trigeminal structures and a state of permanent unconsciousness is induced.

Using the decerebrate preparation, Woodworth and Sherrington (1904) illustrated the essential contribution of the cortex to the perception of pain and defined the "pseudaffective" reflex. In response to a noxious stimulus, this reflex corresponds to a remarkable behavioral repertoire, even including occasional vocalization, due to the fact that its pathways are coordinated at spinal and supraspinal brainstem levels below the midbrain transection (i.e., it is a spino-bulbo-spinal reflex; Woodworth and Sherrington 1904). Despite the behaviors observed, no pain is experienced. In fact, the AVMA

Guidelines on Euthanasia state that "for pain to be experienced, the cerebral cortex and subcortical structures must be functional" (AVMA 2007, p. 2). The pseudaffective reflex is useful in animal studies that investigate neurons of the spinal cord without the influence of anesthesia (e.g., the decerebrate animal preparation). It should be noted that decerebrate preparations are necessarily nonsurvival experiments; Silverman and colleagues (2005, p. 1) note that an animal that recovers from the anesthesia for this procedure typically provides research data "for a period of a few hours or a day" after which it must be euthanized.

Because decerebration severs the connection between the rostral part of the brain and lower CNS structures, it also eliminates the powerful modulatory control mechanisms that descend from supraspinal sites. These descending control mechanisms are predominantly inhibitory and act as a "brake" on spinal cord neurons and circuits that process nociceptive information. Their removal via decerebration leads to enhanced nociceptive reflexes and spinal neuron responses to nociceptive input. Accordingly, spinal cord transection often follows decerebration to enable physiological studies in unanesthetized animals, but it is not a prerequisite of the decerebrate preparation.

Finally, it is important to distinguish the decerebrate from the decorticate preparation. In the latter, only the cerebral cortex is removed, leaving intact the underlying subcortical structures (i.e., the thalamus, brainstem, and spinal cord). Because there have been suggestions that under some conditions pain processing can occur even at the level of the thalamus (e.g., Merker 2007), studies of decorticate animals (which these days are rare) must be performed under general anesthesia.

THE DEVELOPMENT OF PERSISTENT PAIN

The mechanisms that contribute to the development of postoperative/postprocedural and persistent pain are far more complicated than the rather simple anatomical and physiological underpinnings of momentary pain. It is important to appreciate that these types of pain are not merely instances of momentary pain that do not resolve quickly. Rather, they arise in the context and environment of tissue or nerve injury and involve changes in the properties not only of nociceptors but also of the circuits that these receptors engage in the spinal cord and at other levels of the neuraxis (Basbaum and Woolf 1999; Urban and Gebhart 1999; Basbaum and Jessell 2000; Julius and Basbaum 2001). These changes generally enhance signals in "pain" transmission circuits, such that innocuous stimuli can evoke behaviors indicative of pain (extensive discussion of the *sickness syndrome*, an underappreciated postoperative occurrence, is in Chapter 4). As a result of advances in scientific understanding of these mechanisms, many pharmacological treatments for postoperative/-procedural and persistent pain in humans are directed at interfering with the development and duration of hyperalgesia and allodynia.

Hyperalgesia is a hallmark of inflammatory pain and is a consequence of many types of tissue insults (ranging from a skin incision to nerve injury). It is defined as an increased response to a noxious stimulus and manifests as an increased sensitivity to pain (Treede et al. 1992; Campbell and Meyer 2006). Because the threshold for response also typically decreases, sometimes even nonnoxious stimuli can cause pain, a phenomenon called allodynia.

There are two types of hyperalgesia, primary and secondary, each associated with different mechanisms. Primary hyperalgesia is characterized by increased excitability of nociceptors at the site of the insult (e.g., the site of an incision). It occurs most commonly after skin injury, but may also develop following insults to joints, muscle, or viscera. For example, when an incision in the skin is examined, the response to stimuli applied to that site typically increases. Surrounding the site of injury, and often at sites rather distant from the injury (particularly when joints and especially the viscera are involved), is an area of increased sensitivity referred to as the area of secondary hyperalgesia. This is most evident with visceral insult, where sensations are referred or perceived to arise from overlying structures, most notably skin. The classic example is myocardial oxygen deficiency (angina) in which the pain is referred to the shoulder, down the left arm, and occasionally up to the jaw.

When either primary or secondary hyperalgesia occurs, it is accompanied by an increase in the excitability and responses of neurons in the nervous system. Primary hyperalgesia is largely attributed to an increase in the excitability of nociceptors (i.e., the peripheral afferent sensory ending and fiber), whereas secondary hyperalgesia is associated with changes in the excitability of neurons in the CNS, including the spinal cord and supraspinal sites in the brain. Accordingly, primary hyperalgesia is associated with peripheral sensitization of nociceptors and secondary hyperalgesia with central sensitization. The terms indicate an increase in the excitability and responses of peripheral (i.e., nociceptor) and central neurons because of tissue insult.

Numerous mediators in both the peripheral and central nervous systems contribute to the processes of sensitization (Basbaum and Jessell 2000; Basbaum and Woolf 1999; Julius and Basbaum 2001; Treede et al. 1992; see McMahon et al. 2005 for an overview). At the injury site, primary hyperalgesia is induced by the release of numerous inflammatory mediators including the products of cyclooxygenase enzyme activation. The critical contribution of these enzymes accounts for the beneficial effects of nonsteroidal anti-inflammatory drugs, which, by inhibiting the enzyme, reduce peripheral sensitization and help alleviate persistent or postoperative/-procedural pain.

Central sensitization is a considerably more complicated process that can result from changes in the amount of neurotransmitter released from nociceptor terminals in the spinal cord or brainstem, notably glutamate and the neuropeptide substance P (Basbaum and Jessell 2000; Basbaum and Woolf 1999; Woolf 1983); from loss of inhibitory regulation exerted by inhibitory interneurons in the spinal cord and at supraspinal loci; and from biochemical changes in the "pain" transmission neurons that increase their responsiveness to peripheral inputs. It is likely that the pain-alleviating effects of drugs such as ketamine are partly due to the reduction of central sensitization produced by the release of glutamate. In contrast, the beneficial effects of anticonvulsants for pain treatment are likely related to their blockade of neurotransmitter release from primary afferents or the enhancement of inhibitory controls.

The remarkable number of molecules implicated in central sensitization (whether produced by tissue or nerve injury) may lead to the development of new pharmacological approaches to managing persistent pain. Of particular interest is the recent understanding of the contribution of glia to the process of central sensitization. In fact, there is considerable evidence that glia, notably microglia and astrocytes, are activated in the setting of nerve injury and that they are the source of mediators that enhance the central consequences of nociceptor activity (Thacker et al. 2007; Watkins et al. 2007). For this reason, there are now several pharmaceutical programs for the development of novel pain therapies that attempt to interfere with the biochemistry of the "activated" glial cell.

ONTOGENY OF PAIN

Large numbers of developmental neurobiology studies have increased knowledge of the origin and maturation of nociceptive circuitry and behavior. Importantly, it is now possible to identify subpopulations of sensory neurons, including nociceptors, early in embryonic development, well before they project to central and peripheral targets (Fitzgerald 2005).

Neurogenesis and subsequent maturation and synaptogenesis of sensory neurons occur in two waves. In rats, outgrowth of myelinated A-delta fibers from the neuraxis precedes outgrowth of unmyelinated C-fibers. These processes occur during embryonic days 15 to 17 (E15-17) and 18 to 20 (E18-20) respectively and coincide with the first appearance of reflex responses to mechanical stimuli (Fitzgerald 2005). A-delta fiber synapses have been identified in the spinal dorsal horn at E13 in rats, whereas the terminals of C-fibers do not appear until E18-19 (ibid.). In fact, physiological recordings of nociceptive fibers in rat pups during the first few postnatal days demonstrate responses to noxious chemical, mechanical, and thermal stimuli that are similar to those of mature C-fibers.

Neonates of multiple species demonstrate exaggerated spinally mediated reflex responses to noxious stimuli compared to adults (see Fitzgerald 2005 and Hathway and Fitzgerald 2008 for reviews). For example, in rat pups it is not until postnatal day 10 (P10) that these reflexes develop spatial precision; they then achieve adult levels of both spatial and temporal precision by P21. Nonnoxious tactile stimuli are important for fine-tuning of nociceptive reflexes during this critical postnatal period. Likewise, maturation of ascending and descending neuronal pathways, at approximately P10 in rat pups, contributes to the development of mature nociceptive processing. Hyperalgesia can be documented in rat pups as young as 3 days of age, but it is significantly less prominent, both in magnitude and duration, at early ages than it is in the adult animal. By approximately 34 to 40 days of age, adult-like hyperalgesia is evident (Jiang and Gebhart 1998). Taken together, these observations demonstrate the maturation of synaptic connections in the superficial laminae of the dorsal horn during the first 3 postnatal weeks (Fitzgerald 2005).

Both somatic and visceral tissue insults in the neonate appear to alter processing of nociceptive inputs in adulthood. Neonatal injury has thus been associated with either hyperalgesia or hypoalgesia, depending on the type and severity of injury and the sensory modality tested (Bhutta et al. 2001). Colorectal distension in neonatal rats (P8-12) results in colon hypersensitivity in adults (Al-Chaer et al. 2000). In addition to altered nociceptive processing, repetitive or persistent pain in the neonatal period leads to changes in brain development, widespread alterations in animal behavior, and increased vulnerability to stress and anxiety disorders or chronic pain syndromes (Anand et al. 1999, 2007; Al-Chaer et al. 2000; Bhutta et al. 2001).

Specifically, inflammation produced by repeated injections of complete Freund's adjuvant in rat pups (P0, P3, P14) leads to hyperalgesia and lasting changes in nociceptive circuitry of the adult dorsal horn (Ruda et al. 2000). Similarly, rat pups that received repeated formalin injections in the paw developed generalized thermal hypoalgesia as they aged (Bhutta et al. 2001). When noxious formalin stimuli were preceded by morphine analgesia in neonatal rats, hyperalgesia in adulthood was significantly reduced (ibid.). In other models of persisent pain, rat pups less than 21 days old did not develop signs of neuropathic pain after nerve injury (Howard et al. 2005).

Whereas a growing number of studies have demonstrated altered pain processing after neonatal injury in humans, not all outcomes reported are necessarily applicable to the laboratory animal (e.g., see Grunau and Tu 2007). However, an important conclusion from this body of research is that untreated neonatal pain can permanently alter sensitivity to pain, consistent with modulation of primary afferent activation and central sensitization in response to subsequent nociceptive challenges in adulthood. Thus measures to minimize pain in neonates may reduce alterations in neuronal development and long-term sensitivity to sensory stimuli.

MODULATORY INFLUENCES ON PAIN: ANXIETY, FEAR, AND STRESS

As noted above, pain is not merely the appreciation of the presence, location, and magnitude of nociceptive input but rather a complex event with an important emotional/affective component. In addition, psychological factors can significantly influence the experience of pain (also discussed in Chapter 1, text and Figure 1-1). For example, fear and anxiety can enhance responses to and interpretation of pain-producing events (Hunt and Mantyh 2001; Linton 2000; Morley 1999; Munro 2007; Perkins and Kehlet 2000; Ploghaus et al. 2001). For these reasons, the predisposition of certain strains of animals or individuals to anxiety should be considered in efforts to assess the possible contribution of anxiety to the experience of pain (Ulrich-Lai et al. 2006). In humans, measures to reduce anxiety can reduce pain—this is true for both behavioral (cognitive) interventions and anxiolytic drugs (Belzung 2001). Similarly, behavioral interventions to reduce anxiety in animals can include acclimation to human handlers, training to withstand some research procedures, socialization and housing with cage mates, or training and exercise. Reliable and reproducible testing of animals is best achieved in a situation in which the animal is habituated to the test apparatus and the test environment (e.g., light, noise, temperature, humidity).

The extent to which stress is present in normal laboratory situations should also be considered. There are numerous examples in which exposure to stressors can influence the response to a noxious stimulus. Somewhat paradoxically, the response can manifest as an apparent reduction of pain, a phenomenon referred to as "stress-induced analgesia" (Amit and Galina 1986; Keogh and Cochrane 2002; for commentary on how exposure to a predator reduces nociceptive responses in rats see Lester and Fanselow 1985). Moreover, environmental enrichment may also affect stress-related nociceptive responses. A recent study reported that C3H mice exposed to environmental enrichment, which can reduce stress compared with a standard environment (i.e., standard plastic cages with bedding), reacted more quickly (i.e., exhibited a shorter freezing time) to electric shock training than did mice habituated in standard housing conditions. Such an outcome, possibly due to decreased fearfulness or anxiety, may require more nuanced staff training in recognizing modulatory influences on painful situations (Benaroya-Milshtein et al. 2004).

Whether the magnitude of stress experienced in typical laboratory settings is sufficient to significantly alter the perception of pain is difficult to determine. A priori one would assume that reducing stress is a good objective for both experimental outcomes and animal welfare, since perturbation of the latter may lead to stress/distress (see the 2008 NRC report *Recognition and Alleviation of Distress in Laboratory Animals* for detailed information on the effects of stress/distress on animal welfare). The stressors typically used to evoke stress-induced analgesia are intense and rather unnatural and can be useful for evaluating pain behavior in response to an applied stimulus. How data from such studies translate into the normal behavioral repertoire of animals in a laboratory environment and in other types of experimental studies remains to be determined. Nevertheless, it is important to keep in mind the possibility of stress-induced effects when assessing pain in animals because the absence of response to a noxious stimulus or of pain-indicative behavior may be due to significant stress and misleadingly suggest the absence of pain. Because pain can be enhanced by anxiety or fear, readers should consult the discussion of the role of anxiolytics in pain management in Chapter 4.

CONCLUSIONS AND RECOMMENDATIONS

Pain is not a foregone outcome when an animal is exposed to a noxious stimulus, because, as discussed in Chapter 1, the experience of pain is informed by the perceptive abilities of the brain.

1. It is critical to appreciate that nociception is not equivalent to pain. Noxious stimuli trigger several levels of information processing as the activity of primary afferent nociceptors is conveyed to the spinal cord and from there to the higher centers of the brain. Neurons at many levels of the neuraxis respond to noxious stimuli, but that response does not necessarily indicate or lead to pain. In fact, studies of animals with transections of the neuraxis at various levels illustrate that complex responses can be elicited in the absence of pain (i.e., when the cortex is disconnected from the nociceptive processing networks).
2. Until better methods (e.g., biomarkers, imaging) are available to objectively measure pain, behavioral indices and to some extent extrapolation from the human experience are the best sources of information and the only methods available to assess pain in laboratory animals (see Chapter 3).
3. Pain is not exclusively associated with noxious stimuli. After some injuries (e.g., nerve injury), even innocuous stimuli can cause pain, and repeated exposure to noxious stimuli can lead to sensitization and enhance responses to subsequent stimuli both innocuous and noxious.
4. Injury may have long-term consequences to the neural systems

that process nociceptive information. This is particularly true of procedures performed in the neonatal animal, but it may also be relevant in the adult. This information underscores the importance of adequate postoperative pain management and to some extent provides the rationale for preemptive analgesia (see Chapter 4). Psychological factors also likely contribute to the pain experienced during and after an injury; their effect is perhaps more difficult to assess and address in the context of laboratory experiments, but its recognition is important.

5. Pain represents a cascade of physiological, immunological, cognitive, and behavioral effects that may confound experimental results in addition to being detrimental to the animals' welfare.

Finally, and as discussed in Chapter 1, unless not recommended due to experimental outcomes, relief from pain is an ethical and regulatory obligation. Further, the committee emphasizes that effective pain management is scientifically advantageous, as unalleviated pain may adversely influence scientific projects and research outcomes in a number of ways. The reader is referred to Box 1-4 of Chapter 1 and to Chapter 4 for an extended discussion of the consequences of unrelieved pain.

REFERENCES

Al-Chaer ED, Kawasaki M, Pasricha PJ. 2000. A new model of chronic visceral hypersensitivity in adult rats induced by colon irritation during postnatal development. Gastroenterology 119(5):1276-1285.

Amit Z, Galina ZH. 1986. Stress-induced analgesia: Adaptive pain suppression. Physiol Rev 66(4):1091-1120.

Anand KJ, Coskun V, Thrivikraman KV, Nemeroff CB, Plotsky PM. 1999. Long-term behavioral effects of repetitive pain in neonatal rat pups. Physiol Behav 66:627-637.

Anand KJ, Garg S, Rovnaghi CR, Narsinghani U, Bhutta AT, Hall RW. 2007. Ketamine reduces the cell death following inflammatory pain in newborn rat brain. Pediatr Res 62:283-90.

Apkarian AV, Bushnell MC, Treede RD, Zubieta JK. 2005. Human brain mechanisms of pain perception and regulation in health and disease. Eur J Pain 9(4):463-484.

AVMA (American Veterinary Medical Association). 2007. AVMA Guidelines on Euthanasia. Available at www.avma.org/issues/animal_welfare/euthanasia.pdf. Accessed June 9, 2008.

Basbaum A, Jessell T. 2000. The perception of pain. In: Kandel ER, Schwartz JH, Jessell TM, eds. Principles of Neural Science, 4th ed. New York: McGraw-Hill, Health Professions Division.

Basbaum AI, Woolf CJ. 1999. Pain. Curr Biol 9:R429-R431.

Belzung C. 2001. The genetic basis of the pharmacological effects of anxiolytics: A review based on rodent models. Behav Pharmacol 12(6-7):451-460.

Benaroya-Milshtein N, Hollander N, Apter A, Kukulansky T, Raz N, Wilf A, Yaniv I, Pick CG. 2004. Environmental enrichment in mice decreases anxiety, attenuates stress responses and enhances natural killer cell activity. Eur J Neurosci 20(5):1341-1347.

Bhutta AT, Rovnaghi C, Simpson PM, Gossett JM, Scalzo FM, Anand KJ. 2001. Interactions of inflammatory pain and morphine in infant rats: Long-term behavioral effects. Physiol Behav 73(1-2):51-58.

Bingel U, Tracey I. 2008. Imaging CNS modulation of pain in humans. Physiology 23(6): 371-380.

Campbell JN, Meyer RA. 2006. Mechanisms of neuropathic pain. Neuron 52(1):77-92.

Fitzgerald M. 2005. The development of nociceptive circuits. Nat Rev Neurosci 6(7): 507-520.

Grunau RE, Tu MT. 2007. Long-term consequences of pain in human neonates. In: Anand KJ, McGrath PJ, Stevens B, eds. Pain in Neonates and Infants, 3rd Ed. Pain Research and Clinical Management. Amsterdam: Elsevier. pp. 45-55.

Hathway GJ, Fitzgerald MF. 2008. The development of nociceptive systems. In: Bushnell MC, Basbaum AI, eds. The Senses: A Comprehensive Reference, Vol. 5. Pain. San Diego: Academic Press. pp. 133-145.

Howard RF, Walker SM, Mota PM, Fitzgerald M. 2005. The ontogeny of neuropathic pain: Postnatal onset of mechanical allodynia in rat spared nerve injury (SNI) and chronic constriction injury (CCI) models. Pain 115(3):382-389.

Hunt SP, Mantyh PW. 2001. The molecular dynamics of pain control. Nat Rev Neurosci 2(2):83-91.

Jiang MC, Gebhart GF. 1998. Development of mustard oil-induced hyperalgesia in rats. Pain 77(3):305-313.

Julius D, Basbaum AI. 2001. Molecular mechanisms of nociception. Nature 413:203-210.

Keogh E, Cochrane M. 2002. Anxiety sensitivity, cognitive biases, and the experience of pain. J Pain 3(4):320-329.

Lester LS, Fanselow MS. 1985. Exposure to a cat produces opioid analgesia in rats. Behav Neurosci 99(4):756-759.

Linton SJ. 2000. A review of psychological risk factors in back and neck pain. Spine 25(9): 1148-1156.

McMahon SB, Bennett DL, Bevan S. 2005. Inflammatory mediators and modulators of pain. In: Koltzenburg M, McMahon S, eds. Wall and Melzack's Textbook of Pain 5th ed. Churchill-Livingstone. pp. 49-72.

Merker B. 2007. Consciousness without a cerebral cortex: A challenge for neuroscience and medicine. Behav Brain Sci 30(1):63-81; discussion 81-134.

Morley JS. 1999. New perspectives in our use of opioids. Pain Forum 8(4):200-205.

Munro G. 2007. Dopamine D(1) and D(2) receptor agonism enhances antinociception mediated by the serotonin and noradrenaline reuptake inhibitor duloxetine in the rat formalin test. Eur J Pharmacol 575(1-3):66-74.

NRC (National Research Council). 2008. Recognition and Alleviation of Distress in Laboratory Animals. Washington: National Academies Press.

Perkins FM, Kehlet H. 2000. Chronic pain as an outcome of surgery: A review of predictive factors. Anesthesiology 93(4):1123-1133.

Ploghaus A, Narain C, Beckmann CF, Clare S, Bantick S, Wise R, Matthews PM, Rawlins JN, Tracey I. 2001. Exacerbation of pain by anxiety is associated with activity in a hippocampal network. J Neurosci 21(24):9896-9903.

Ruda MA, Ling QD, Hohmann AG, Peng YB, Tachibana T. 2000. Altered nociceptive neuronal circuits after neonatal peripheral inflammation. Science 289(5479):628-631.

Sherrington CS. 1906. The Integrative Action of the Nervous System. New York: Charles Scribner's Sons.

Silverman J, Garnett NL, Giszter SF, Heckman CJ II, Kulpa-Eddy JA, Lemay MA, Perry CK, Pinter M. 2005. Decerebrate mammalian preparations: Unalleviated or fully alleviated pain? A review and opinion. Cont Top Lab Anim 44(4):34-36.

Thacker MA, Clark AK, Marchand F, McMahon SB. 2007. Pathophysiology of peripheral neuropathic pain: Immune cells and molecules. Anesth Analg 105(3):838-847.

Tracey I, Mantyh PW. 2007. The cerebral signature for pain perception and its modulation. Neuron 55(3):377-391.

Treede RD, Meyer RA, Raja SN, Campbell JN. 1992. Peripheral and central mechanisms of cutaneous hyperalgesia. Prog Neurobiol 38(4):397-421.

Ulrich-Lai YM, Xie W, Meij JT, Dolgas CM, Yu L, Herman JP. 2006. Limbic and HPA axis function in an animal model of chronic neuropathic pain. Physiol Behav 88(1-2):67-76.

Urban MO, Gebhart GF. 1999. Supraspinal contributions to hyperalgesia. Proc Natl Acad Sci U S A 96:7687-7692.

Watkins LR, Hutchinson MR, Milligan ED, Maier SF. 2007. "Listening" and "talking" to neurons: Implications of immune activation for pain control and increasing the efficacy of opioids. Brain Res Rev 56(1):148-169.

Woodworth RS, Sherrington CS. 1904. A pseudoaffective reflex and its spinal path. J Physiol (Lond) 31:234-243.

Woolf CJ. 1983. Evidence for a central component of post-injury pain hypersensitivity. Nature 306(5944):686-688.

3

Recognition and Assessment of Pain

This chapter begins with a presentation of the clinical signs and behaviors that veterinarians use to recognize animals in pain. It then provides a review of methods for pain assessment, with a focus on techniques for specific laboratory animal species. It concludes with species-specific clinical signs and behavioral responses to pain.

INTRODUCTION

Recognizing pain and assessing its intensity are both essential for its effective management. If pain is not recognized, then it is unlikely to be treated; failure to appreciate the intensity of pain will hamper the selection of an appropriately potent analgesic, raise doubts about the effectiveness of the administered dose, and result in less than optimal treatment. In humans, self-report of pain is the "gold standard" by which other assessment techniques may be judged, although there are limitations and biases even when using this approach (see Chapter 1). For animals, as for humans who cannot self-report (e.g., the very young and those with cognitive impairment; Ranger et al. 2007; Zwakhalen et al. 2006), other assessment tools are necessary.

Since the publication of the first edition of this report (NRC 1992), there have been considerable advances in scientists' understanding of animal pain and numerous attempts to develop methods of assessing pain. Yet few validated assessment techniques are available. In most circumstances pain is assessed based on an animal's clinical appearance and overall behavior. Although this approach can be unreliable, it is usually effective in detecting severe pain in many species. It is also effective when pain is localized to one limb (causing lameness) or to a specific body area (resulting in a marked behavioral response if that area is palpated).

The ability to assess pain in laboratory animals will improve with the development of validated, objective schemes for particular species and types of procedures. Some schemes of this type are in development, while others (e.g., assessment of postsurgical pain in dogs [Morton et al. 2005] or of pain after abdominal surgery in rats [Roughan and Flecknell 2001, 2003]) have reached the point that they can be used to assess pain in the particular species in a variety of situations. It is also possible that some of the behaviors noted may occur in other species: contraction of the abdominal muscles following abdominal surgery is observed in rats and has also been reported in mice (Wright-Williams et al. 2007) and rabbits (Leach et al. 2009). Regardless of the assessment technique, however, it is important that it be done by a team that includes researchers, veterinarians, and animal care staff.

PAIN RECOGNITION: CLINICAL SIGNS AND BEHAVIOR

There are no generally accepted objective criteria for assessing the degree of pain that an animal is experiencing. Species vary widely in their response to pain, and often animals of the same species show different responses to different types of pain. Box 3-1 presents a basic algorithm for pain assessment that may serve until the development of species-specific pain assessment methods. A team approach and cooperative spirit among all interested parties—researchers, veterinarians, and animal care staff—will benefit the welfare of the animal in pain.

It is important that clinical evaluations and assessment protocols be carried out by individuals with a detailed knowledge of the normal and abnormal behavior and appearance of the species concerned. Further, the effects of the observer on the behavior of the animal should be considered; for example, some species, such as rabbits and guinea pigs, may remain immobile, especially if the observer is an unfamiliar person. In these cases, it may be necessary to observe the animal via a camera or viewing panel. When assessing behavioral changes, it is often helpful to have a checklist that may incorporate a grading scheme (see the scoring system developed by Morton and Griffiths in 1985). However, because different individuals often fail to agree on the score that should be assigned (Beynen et al. 1987) it may be simpler to note the presence or absence of a specific clinical sign. Changes in successive observations could indicate an improvement or deterioration in the animal's condition. Although many observations will not be specific indicators of pain, a structured examination is always helpful in monitoring an animal's progress during a study. Table 3-1 presents a number of behavioral signs usually associated with pain.

BOX 3-1
Pain Assessment Protocol

The following approach can be helpful for assessing pain in particular animal models:

- Prepare a checklist of the examinations to be undertaken, allow space for a general comment, and perhaps include an overall assessment tool (e.g., a visual analogue scale (VAS) score sheet). Familiarize all staff who will be involved in the assessment with this checklist and any other assessment tools that will be used. Whenever possible, the same staff member should conduct each assessment of the same animal. Specific training must be provided for new or inexperienced staff.
- Begin by observing the animal without disturbing it. If the animal's behavior changes markedly in the presence of an observer (e.g., as is the case with nonhuman primates, rabbits, and guinea pigs) it may be more practical to assess postoperative or postprocedural behavior by setting up a video camera or viewing panel.
- Assess the animal's response to the observer (the technician who routinely cares for the animal may be best able to assess this).
- Examine the animal and assess its response to gentle palpation or handling of any presumed painful areas (e.g., the site of surgery, the site of a lesion) when practicable.
- Weigh the animal, record its food and water consumption if possible, and examine the cage or pen for signs of normal or abnormal urination or defecation.
- Administer analgesic treatment if necessary, and repeat the assessment outlined above 30-60 minutes after treatment to determine whether the drug and the dose administered have been effective. In the absence of certainty about the presence of pain, assessing the response to an analgesic can be helpful.
- Review these protocols regularly.
- Remember that:
  - the signs described here can be caused by conditions other than pain,
  - the signs may vary between animals of the same species, even after the same procedure, and
  - the signs will vary between different strains and breeds.

Animals in pain reduce their overall level of activity, as observed, for example, in mice following surgery (Clark et al. 2004; Karas 2002; Wright-Williams et al. 2007). It has been suggested that changes in heart rate, respiratory rate, and blood pressure can be used to assess pain, but these clinical parameters are often unreliable or nonspecific (e.g., similar changes may be observed in stressed or distressed animals; NRC 2008). Consistent changes in these parameters in animals expected to be in pain have not been demonstrated (Cambridge et al. 2000; Holton et al. 1998; Price et al. 2003). Given the range of factors (e.g., fear, excitement) that can alter heart and respiratory rate, this is not surprising, as even handling can cause major

TABLE 3-1 Behavioral Signs of Persistent Pain

Sign	Explanation
Guarding	The animal alters its posture to avoid moving or causing contact to a body part, or to avoid the handling of that body area.
Abnormal appearance	Different species show different changes in their external appearance, but obvious lack of grooming, changed posture, and a changed profile of the body are all observable signs. In species capable of some degree of facial expression, the normal expression may be altered.
Altered behavior	Behavior may be depressed; animals may remain immobile, or be reluctant to stand or move even when disturbed. They may also exhibit restlessness (e.g., lying down and getting up, shifting weight, circling, or pacing) or disturbed sleeping patterns. Large animal species may grunt, grind their teeth, flag their tail, stomp, or curl their lips (especially sheep and goats). Primates in pain often roll their eyes. Animals in pain may also show altered social interactions with others in their group.
Vocalization	An animal may vocalize when approached or handled or when a specific body area is touched or palpated. It may also vocalize when moving to avoid being handled.
Mutilation	Animals may lick, bite, scratch, shake, or rub a painful area.
Sweating	In species that sweat (horses), excessive sweating is often associated with some types of pain (e.g., colic).
Inappetence	Animals in pain frequently stop eating and drinking, or markedly reduce their intake, resulting in rapid weight loss.

changes in heart rate, respiratory rate, and blood pressure. Recently, however, more sophisticated analysis of heart rate variability has been of value as an adjunct to pain assessment (Arras et al. 2007; Rietmann et al. 2004).

PAIN ASSESSMENT METHODS

As discussed above, methods for assessing pain in laboratory animals remain highly subjective and are based largely on preconceived ideas about the appearance and behavior of animals in response to pain. Attempts to apply the Morton and Griffiths (1985) scoring scheme were largely unsuccessful (Beynen et al. 1987), primarily because the variables selected for inclusion were not fully identified and the ratings (0-3) not sufficiently well characterized (this scheme has proven much more successful in the development of humane endpoints for studies that may cause distress rather than pain; NRC 2008).

In addition to the lack of known effective pain assessment methods, it is not uncommon for a study to include the administration of an analgesic without any attempt to evaluate its effectiveness. For example, a recent survey of pain control in laboratory animals in the United Kingdom found that, although all the institutions in the survey used analgesics, almost none used methods of pain assessment to confirm that the treatment was effective (Hawkins 2002).

Behavioral Changes

Objective measures likely to indicate pain include changes in general locomotor activity (e.g., guarding a specific area or avoiding weight-bearing on an injured limb; Duncan et al. 1991; Flecknell and Liles 1991; Malavasi et al. 2006) and in food and water intake and body weight (Liles and Flecknell 1992, 1993a,b). These measures are also useful to assess analgesic drug efficacy, although because they are retrospective they cannot be used to modify analgesic therapy for a particular animal. They are, however, effective as a simple measure of postoperative recovery and as a means of adjusting future analgesic regimens for similar animals undergoing similar surgical procedures.

Influences of Analgesics on Behavior

The use of analgesics warrants certain cautions. Some analgesics, notably opioids, cause marked behavioral changes in healthy, pain-free animals, which can confound attempts to assess pain (Roughan and Flecknell 2000). Buprenorphine stimulates activity in normal mice (Cowan et al. 1977; Hayes et al. 2000), so behavioral changes after the use of this drug during surgery could be due to the provision of effective pain relief or a nonspecific drug effect. In contrast, NSAIDs have only very minor effects on behavior in healthy, pain-free animals, so this problem is not significant with the use of these analgesics (Roughan and Flecknell 2001; Wright-Williams et al. 2007).

Further, significant behavioral signs of postsurgical pain in rodents may persist only 6 to 8 hours after some procedures (Roughan and Flecknell 2004), so these results may be due to administration of analgesics to animals that were not experiencing pain. In these circumstances side effects such as sedation or nausea may be of much greater significance. For more information on other behavioral measures readers are referred to Chapter 1, especially Box 1-4.

Moreover, the influence of analgesics on body weight following surgery is not always easy to interpret. In some studies, after an initial presumed beneficial effect, animals that had undergone surgery and not received postoperative analgesics gained more weight over a 2- to 3-day period than their counterparts under an analgesic regime (Sharp et al. 2003).

Behavioral Assessment Studies in Rats, Mice, and Rabbits

Investigators have described specific behavioral changes following abdominal surgery and ureteral calculosis in rats (Giamberardino et al. 1995; Gonzalez et al. 2000; Roughan and Flecknell 2000) and these behaviors have been incorporated in a practicable pain assessment tool for use in laboratory rats after abdominal surgery (Roughan and Flecknell 2002). During the initial development of the scheme, rat behavior was evaluated both before and after a midline laparotomy with appropriate untreated and anesthetic and analgesic controls.

An initial study using buprenorphine as the analgesic was inconclusive because of the marked effects of this opioid on normal behavior (Roughan and Flecknell 2001). A subsequent study using carprofen and ketoprofen successfully identified behaviors that differentiated rats that had (1) undergone surgery from those that had simply been anesthetized and (2) received analgesics after surgery from those that had not. These studies required detailed analysis of considerable periods of videotaped behavior including filming at night under red light. The utility of these behaviors was further demonstrated in rats undergoing surgery as part of an unrelated research project that entailed placing the animals in an observation cage for a 15-minute period and assessing the frequency of the pain-related behaviors. Again, it was possible to differentiate animals receiving analgesics from untreated controls, and to demonstrate a dose-related effect of the NSAID meloxicam (Roughan and Flecknell 2003).

When experienced staff (animal technicians, research workers, and veterinarians) first viewed selected video recordings from these animals, they were unable to correctly identify the treatment groups. However, after watching a short recording illustrating the key behaviors, their ability to identify animals that had or had not received analgesics greatly improved (Roughan and Flecknell 2006). These studies suggest that key behaviors can be identified and used to score pain following one type of surgical procedure in rats. In addition, the studies underscore the importance of proper training of even experienced personnel with the introducton of new techniques. It is not yet clear whether behavioral changes in rats after various surgical procedures will differ greatly in type or will be drawn from a common group of abnormal, pain-related behaviors.

Recent studies in mice have indicated that they experience similar pain-related changes in behavior after abdominal surgery (Wright-Williams et al. 2007) and that these behaviors might form the basis of a murine pain scoring scheme. However, the rapid movement of mice makes observations less reliable. In addition, the effects of the analgesics used in these studies were less predictable than in rats as were the effects of opioids, which, as mentioned above, affect behavior in normal animals. These studies also found a major difference in the frequency of pain-related behaviors in the two different strains of mice used (C3He and C57Bl6). Other studies (e.g., Karas 2002) have shown changes in the frequency of normal activity in mice after surgery, and it may be possible to develop a scoring system based on a combination of changes in abnormal and normal activity.

In some instances, changes in a specific locomotor pattern, or gait, can be assessed objectively using a variety of techniques (Gabriel et al. 2007). Force plates and other means of assessing limb use and gait have been used to evaluate the severity of arthritis in laboratory and companion animals as well as the efficacy of analgesic therapy (Gabriel et al. 2007; Hazewinkel et al. 2008). The linking of clinical signs to behavioral alterations after administration of an analgesic facilitates pain assessment.

A small number of studies have attempted to assess postsurgical pain in rabbits. Initial attempts to develop a behavior-based scheme failed because of the animals' reaction to the presence of an observer (Roughan and Flecknell 2004), and a similar study produced inconclusive results (Parga 2002). More recently, a detailed assessment of behavior before and after surgery, using remotely operated cameras, revealed clearly identifiable abnormal behaviors as well as changes in the frequency of normal behaviors. The effects of analgesics were limited. Further work is required before clear recommendations can be made about the usefulness of these behaviors (Leach et al. 2009).

A problem with all of these behavior-based schemes is that in many instances the animals studied were anesthetized with regimens (e.g., isoflurane or sevoflurane) that resulted in rapid recovery of consciousness. When recovery is delayed, or is associated with prolonged sedation, animals may fail to express pain behavior and scoring may therefore not be reliable. The scoring system may also be influenced by other factors, such as the animals' fear and apprehension, or unexpected variations in behavior between different strains (Wright-Williams et al. 2007). Nevertheless, detailed behavioral observations are a step forward in developing a practical and useful pain scoring system for use after surgery in laboratory animals. What is not yet known is whether similar systems can be used to develop a means of identifying and quantifying other types of pain in animals, including chronic pain.

Developing Objective Pain Assessment Tools: Companion Animals

Initial methods for scoring pain in companion animals were largely subjective and seriously flawed. Some studies, however, demonstrated that behavioral assessments could be used to evaluate the effects of surgery and analgesia (e.g., the use of visual analogue scores to assess pain following ovariohysterectomy in dogs [Lascelles et al. 1997] and cats [Slingsby and

Waterman-Pearson 1998]). Additional scoring schemes for use in dogs have since been developed (Firth and Haldane 1999; Holton et al. 2001), and numerous studies use VAS, numerical rating systems, simple descriptive scores, or a mix of the three approaches (Brodbelt et al. 1997; Mathews et al. 2001). These different approaches highlight the problems involved in developing pain assessment schemes (Holton et al. 1998); for example,

- the assessment criteria are frequently highly subjective,
- the study designs do not include untreated (surgery and no analgesia) controls,
- the study designs do not include anesthesia and analgesia (and no surgery) control groups, and/or
- only a single dosage is assessed rather than a range of doses.

Firth and Haldane (1999) videotaped dogs before and after surgery and, after making detailed observations, identified behaviors that were probable indicators of pain. In common with other behavior-based scoring schemes, they hypothesized that behaviors that appeared only after surgery, or that increased or decreased greatly after surgery, could be pain-related. If administration of an analgesic normalized these behavioral changes, this provided additional evidence that the changes were due to pain. The scheme set out by Firth and Haldane has been developed further and recommended as a tool suitable for clinical use (Gaynor and Muir 2002).

Holton and colleagues (2001) adopted a different approach. This group sought to identify descriptors of pain by consulting with experienced small animal clinicians, and then used sophisticated analytical techniques to reduce these descriptors to a set of words or phrases that could be developed into a multidimensional pain scale. Unfortunately, validation in a placebo-controlled, blinded study has yet to be completed.

It is important to note that the development of a pain score essentially based on the opinion of clinician experts is almost certain to result in a self-fulfilling scheme that will detect pain and predict which animals will receive additional analgesics, since it will be used by clinicians whose opinion shaped its development. Because this is a common problem in pain scoring of both animals and humans, these schemes should be developed further and validated through randomized, blind, placebo-controlled trials.

Placebo-controlled studies in animals, however, pose significant ethical and practical difficulties. Because most schemes include some behavioral assessments, and because anesthetics and analgesics, notably opioids, can markedly change behavior in normal, pain-free animals, lack of appropriate controls (i.e., postprocedural animals that receive no anesthetic or analgesic) can make the results highly questionable. The inclusion, however, of such control groups may cause significant ethical dilemmas to researchers that undertake pain assessment studies, most of which are carried out in veterinary schools. Deliberately withholding analgesics in circumstances believed likely to result in pain may be considered unacceptable by students who learn that animals experience pain and should receive analgesics. Studies of pain with human participants require an intervention analgesia protocol so that subjects assessed as experiencing pain above a predetermined level are removed from the study and given an analgesic. This approach has been used in a number of veterinary clinical studies (Grisneaux et al. 1999; Lascelles et al. 1995).

Measurement of Nociceptive Responses

A wide variety of methods for measuring nociceptive response apply to either momentary or more longer-lasting noxious stimuli for research purposes (Hogan 2002; Le Bars et al. 2001).[1] Although these have limited application for assessing pain in other situations (e.g., after surgery), they do provide insight into potential pain-related behaviors and can help predict effective analgesic drug dose rates. Techniques that measure momentary nociceptive responses involve the application of a brief noxious stimulus followed by quantification of the animal's response. Administration of analgesics usually modifies this response, for example by prolonging the latency of withdrawal of a limb or tail from the noxious stimulus. In addition to the use of such techniques in small laboratory animals, they have been applied to studies in larger species to assess analgesic efficacy and detect the occurrence of hyperalgesia after injury (Dixon et al. 2002; KuKanich et al. 2005; Ley and Waterman 1996; Pypendop et al. 2006; Slingsby et al. 2001; Veissier et al. 2000; Welsh and Nolan 1995).

Although primarily used as a means of screening for potential analgesics in drug discovery programs, the results of nociception measurement have been used to estimate dose rates of analgesics for clinical use in both large and small animals. Such extrapolations, however, must be made with caution. In one study, estimates of appropriate doses of buprenorphine based on tail flick tests resulted in a recommended dose of 0.5 mg/kg in rats (Flecknell 1984), 10 times higher than that proven to be effective in postoperative pain scoring systems (Roughan and Flecknell 2004). Since high doses of this agent can have undesirable side effects, it is important to approach these extrapolations very carefully.

Although the results of these tests may not predict clinical efficacy, they

[1]The committee acknowledges the publication of pertinent work on both small laboratory rodents and larger animal species. Readers who wish to delve into this topic are urged to begin with the cited references and expand their reading through them.

do illustrate the very wide variation in response among different strains of rodents (Mogil et al. 1999; Morgan et al. 1999) and thus reinforce the importance of developing pain scoring schemes. If appropriate schemes cannot be used, then dose rates are probably best estimated based on the results of inflammatory pain models such as the late-phase formalin test (Roughan and Flecknell 2002; Appendix A provides details).

Biological Markers of Nociceptor Activation

Although biomarkers of nociceptor activation can be used only as research tools, they can indicate whether a particular procedure could cause pain. For example, the early gene product c-*fos* (Coggeshall 2005) has been used as a marker of nociceptor activity in a number of species (Lykkegaard et al. 2005; Svendsen et al. 2007). Such assessments are possible only within a short time after the animal is euthanized and so are not suitable for routine clinical use.

As discussed in Chapter 2, nociceptor activation and some of the other peripheral and central changes associated with pain and tissue damage result in alterations of sensory thresholds, notably hyperalgesia and allodynia (the perception of previously nonnoxious stimuli as noxious). These changes have been used as indicators of both nociceptor activity and the efficacy of analgesic therapy in both laboratory and clinical studies (Lascelles et al. 1997; Whiteside et al. 2004). Although these methods essentially measure peripheral changes, it is reasonable to assume that in conscious animals such changes indicate that pain has been experienced and may still be present.

Brain Activity Imaging

Recent imaging studies have demonstrated that exposure to noxious stimuli activates a range of cortical and subcortical areas—both primary somatosensory cortex and areas associated with the affective component of pain in humans (Hess et al. 2007). Although such activation does not demonstrate *awareness* of pain in animals, it clearly indicates activation of the cortical areas considered necessary for the affective component of pain (see also Box 1-3). The use of imaging offers a novel approach for detecting central processing of nociceptive information in animals and may enable a more objective assessment of the potential for particular procedures or conditions to cause pain.

PAIN ASSESSMENT: SPECIES-SPECIFIC CLINICAL SIGNS

There is a remarkable lack of validated behavioral signs of pain in many species (Viñuela-Fernández et al. 2007). The following sections present a number of species-specific clinical manifestations based on expert clinical opinion and best practices. Although the signs described typically accompany or indicate pain, many are not specific to pain and may occur as general signs of ill health or as responses to stress or distress (readers are encouraged to consult the ethograms and tables with species-specific clinical signs indicating pain, distress, or discomfort in the appendix of the 2008 NRC report *Recognition and Alleviation of Distress in Laboratory Animals*).

Nonhuman Primates

Nonhuman primates show remarkably little reaction to surgical procedures or to injury, especially in the presence of humans, and might look well until they are gravely ill or in severe pain. Viewing an animal from a distance or by video can aid in detecting subtle clinical changes. A nonhuman primate that appears sick is likely to be critically ill and might require rapid attention.

A nonhuman primate in pain has a general appearance of misery and dejection. It might huddle in a crouched posture with its arms across its chest and its head forward with a "sad" facial expression or a grimace and glassy eyes. It might moan or scream,[2] avoid its companions, and stop grooming. A monkey in pain can also attract altered attention from its cagemates, varying from a lack of social grooming to attack. The animal may show acute abdominal pain through facial contortions, clenching of teeth, restlessness, and shaking accompanied by grunts and moans. Head pain may be manifest by head pressing against the enclosure surface. Self-directed injurious behavior may be a sign of more intense pain. Primates in pain usually refuse food and water. If an animal is well socialized or trained to perform tasks as part of a research protocol, changes in response to familiar personnel or in willingness to cooperate may indicate pain.

Dogs

Dogs in pain generally appear less alert and quieter than normal although small breeds are generally more reactive to environmental changes than large dogs. Dogs in pain may move stiffly or be unwilling to move, and if in severe pain may lie still or adopt an abnormal posture to minimize discomfort. In less severe pain, dogs can appear restless and more alert. Other apparent potential changes include inappetence, shivering, and increased respiration with panting. Dogs in pain may bite, scratch, or guard painful regions and if handled may be unusually apprehensive or aggressive. Their response to a familiar handler may be differ-

[2]Loud and persistent vocalization is an occasional but unreliable expression of pain as it is more likely to signify alarm or anger.

ent; for example, a dog in pain may fail to wag its tail or may shrink away. Incessant licking is sometimes associated with localized pain. Pain in one limb usually results in limping or holding up of the affected limb with no attempt to use it. Spontaneous barking is unlikely; dogs are more likely to whimper or howl, especially if unattended, and may growl without apparent provocation. However, lack of vocalization or excessive vocalization is not a reliable indicator of pain.

Cats

With cats, which are less noticeably reactive to environmental changes than dogs, a general lack of well-being is an important indication of pain. A cat in pain is generally quiet and has an apprehensive facial expression (e.g., its forehead may appear creased). The animal may cry, yowl, growl, or hiss if approached or made to move. It tends to hide or to separate itself from other cats. Its posture becomes stiff and abnormal, varying with the site of pain. If the pain is in the head or ears, the animal might tilt its head toward the affected side. A cat with generalized pain in both the thorax and abdomen may crouch or hunch. If the pain is only thoracic, the head, neck, and body might be extended. A cat with abdominal or back pain might stand or lie on its side with its back arched or walk with a stilted gait. Incessant licking is sometimes associated with localized pain. Pain in one limb usually results in limping or holding up of the affected limb with no attempt to use it. Cats in severe or chronic pain look ungroomed and behave markedly differently from normal. Touching or palpation of a painful area might produce an immediate violent reaction and an attempt to escape. A reduction in food and water intake may be an indicator of pain.

Laboratory Rodents

Rats and mice are the two rodent species most widely used in research generally and in pain-related studies specifically, so it is important that researchers and institutional animal care and use committees recognize when these animals are in pain (for additional information see Chapter 1: Boxes 1-3 and 1-4, Chapter 4, and Appendix A). Rats and mice in acute pain may vocalize and become unusually aggressive when handled. Because rodents also vocalize at ultrasonic frequencies inaudible to humans, the absence of audible vocalization does not necessarily signify the absence of acute pain. Inappetence or a change in feeding activity may become evident; for example, the animals may eat bedding or their offspring. If they are housed with others, the normal group behavior or grooming might change. Rodents in pain may separate from their cagemates and attempt to hide, or they may no longer exhibit nest-building behavior. In rats, porphyrin secretion ("red tears") may appear around the eyes and nose, although this is a general response to stress of any kind.

Normal guinea pigs stampede and squeal when startled, when attempts are made to handle them, or when strangers are in the room, but sick guinea pigs and those in pain usually remain quiet. However, because a normal guinea pig's initial response to the presence of an observer is also to remain immobile, assessing signs of pain can be extremely difficult. Guinea pigs in pain reduce their food and water consumption and may become anorexic. As with rabbits, this behavior can exacerbate the ileus (i.e., gut stasis) that may occur following surgery and can result in a fatal enterotoxemia.

There is virtually no information about signs of pain in hamsters and gerbils, although it is assumed that when in pain they, like rats and mice, will show decreased activity, piloerection, and an ungroomed appearance. As with other species they may adopt an abnormal posture, which may be particularly obvious when moving. Respiration may change.

Rabbits

Rabbits in pain may appear apprehensive, anxious, dull, or inactive, assume a hunched appearance, attempt to hide, and squeal or cry. But sometimes they show aggressive behavior with increased activity and excessive scratching and licking. Reactions to handling are exaggerated, and acute pain might result in vocalization. With abdominal pain, they may show back arching when moving, contraction of the abdominal muscles, and pressing of the abdomen to the ground. Although teeth-grinding has been identified as an indicator of pain, it is not a reliable behavioral sign and studies to support its usefulness as a pain indicator have not yet been done. The respiratory rate of the animals may increase and they may eat and drink less. As with rodents, surgery in rabbits can result in ileus and this, coupled with pain-associated inappetence, can lead to the development of a fatal enterotoxemia. As with other species, a general lack of grooming may be associated with pain.

Ungulates

The greatest progress in developing objective behavior-based methods of assessing the response to pain and injury has been in farm animals. Behavioral and endocrine indicators of pain in lambs, cattle, and pigs have been established by a number of research groups (Hay et al. 2003; Lester et al. 1996; Mellor and Stafford 2000; Molony et al. 2002; Noonan et al. 1994) and vocalization patterns in piglets have been analyzed as potential indicators of pain (Puppe et al. 2005; Weary et al. 1998). These measures have been developed largely to aid in the evaluation of the welfare benefits of modifying standard agricultural practices such as tail docking, castration, and dehorning. It has been repeatedly demonstrated that use of local anesthetics, either alone or in conjunction with modifications to the techniques commonly used, can reduce pain-related behaviors in lambs and cattle (Mellor and Stafford 2000). These studies not only allowed ranking of the degree of pain caused by different procedures but also highlighted some of the problems associated with the use of behavioral signs as indicators of pain. For example, lambs castrated using a rubber ring to constrict the neck of the scrotum show a series of very easily identified abnormal behaviors associated with pain. Lambs castrated surgically without anesthesia remain largely immobile for prolonged periods but the endocrine stress response produced by this method is even greater than that produced by rubber ring occlusion (Lester et al. 1991). Because the types of behaviors observed in lambs undergoing these different procedures varied, it was not possible to use behavior alone to rank the degree of pain. However, the behavioral responses could be used to compare methods of reducing the pain associated with each procedure (Molony et al. 2002).

Horses

Horses in acute pain show reluctance to be handled and other varied responses (Ashley et al. 2005; Driessen and Zarucco 2007): periods of restlessness, interrupted feeding with food held in the mouth uneaten, anxious appearance with dilated pupils and glassy eyes, increased respiration and pulse rate with flared nostrils, profuse sweating, and a rigid stance. Horses in pain also grind their teeth, switch their tails, or play with their water bucket. For animals in prolonged pain, behavior may change from restlessness to depression with head lowered. In pain associated with skeletal damage, there is reluctance to move; the animal may hold its limbs in unusual positions (e.g., it may stand "parked" with the weight on the hind feet and one front foot "pointed" ahead of the other), and the head and neck in a fixed position. Horses with abdominal or thoracic pain may look at, bite, or kick their abdomen; get up and lie down frequently; walk in circles; stand "parked" with elbows adducted; and sweat, roll, and injure themselves as a result of these activities, with bruising especially around the eyes.

Cattle

Cattle in pain often appear dull and depressed, hold their heads low, and show little interest in their surroundings. Their overall activity may be reduced (Hudson et al. 2008). Other observable changes include inappetence, weight loss, grunting, grinding of teeth, and, in milking cows, decreased milk yield (Hernandez et al. 2002, 2005). Severe pain often results in rapid, shallow respiration. On handling, the animals may react violently or adopt a rigid posture to immobilize the painful region. Localized pain may be associated with persistent licking or kicking at the offending area and, when the pain is severe, bellowing. Generally, signs of abdominal pain are similar to those in horses, but less marked. Rigid posture can lead to a lack of grooming because of an unwillingness to turn the neck. With acute abdominal conditions, such as intestinal strangulation, cattle adopt a characteristic stance with one hind foot placed directly in front of the other.

The behavior of calves after dehorning and castration without anesthesia has been described in detail (Molony et al. 1995; Stafford and Mellor 2005) and includes decreased rumination and feeding and an increased incidence of ear flicking, tail flicking, and head shaking. After castration using a rubber ring, calves showed restlessness, foot stamping/kicking, stretching, and adjustments of posture ("easing quarter"); in contrast, after crushing (Burdizzo) or surgical castration the most marked behavioral change was "statue" standing (Molony et al. 1995).

Sheep and Goats

Signs of pain in sheep and goats are generally similar to those in cattle, but sheep, in particular, tolerate severe injury without overt signs of pain or distress. There is a general reluctance to move, coupled with changes in posture, movement, and facial expression. Pain can also cause cessation of rumination, eating, and drinking, and increased curling of the lips; but, as in other species, these are not reliable indicators of pain. Goats are more likely than cattle to vocalize in response to pain. They may also grind their teeth, have rapid and shallow breathing, change posture frequently, and appear agitated (stamping their feet). Dairy goats quickly decrease production and lose body weight and general body condition. After castration or tail docking, lambs show very characteristic signs of pain by standing and lying repeatedly, wagging their tails, occasionally bleating, and displaying neck extension, dorsal lip curling, kicking, rolling, and hyperventilation (Molony et al. 2002).

Pigs

Pigs in pain might show changes in their overall demeanor, social behavior, gait, and posture as well as an absence of bed making. They may become apathetic and unwilling to move and may hide in bedding if possible. Pigs normally squeal and attempt to escape when handled, and pain can accentuate these reactions or cause adults to become aggressive. Squealing is also characteristic when painful areas are palpated. More moderate pain may simply reduce activity levels and make the animal less responsive to familiar handlers and reluctant to feed or drink (Harvey-Clark et al. 2000; Malavasi et al. 2006).

Birds and Poultry

Birds in pain show escape reactions, vocalization, and excessive movement. Small species struggle less and emit fewer distress calls than large species. Head movements increase in extent and frequency. There may also be an increase in heart and respiratory rates. Birds in chronic pain may exhibit a passive immobility characterized by a crouched posture with closed or partially closed eyes and head drawn toward the body. They may also become inappetent and inactive with a drooping, miserable appearance, holding their wings flat against the body and their neck retracted. There may be reduced perching or birds may remain at the bottom of the cage. When a bird is handled, its escape reaction may be replaced by immobility. Birds with limb pain avoid use of the affected limb and refrain from extension.

Reptiles

Acute pain in reptiles is characterized by flinching and muscle contractions. There might be aversive movements away from the unpleasant stimulus and attempts to bite. Chronic and persistent pain may be associated with inappetence, lethargy, and weight loss, although it is difficult to associate any of these signs of lack of well-being specifically with pain.

Fish

It is difficult to determine the nature of the response to pain in fish or whether their experience is similar to that observed in mammals (ILAR 2009; Rose 2002; Sneddon 2006; see Chapter 1). Although there have been few species-specific studies, there is evidence that fish exhibit a pronounced initial response to injuries or to contact with nociceptive stimuli or chemical algesics (Sneddon 2003; Sneddon et al. 2003a,b; Reilly et al. 2008; Ashley et al. 2009) but their response to chronic stimuli has not been characterized. Generally, fish react to noxious stimuli (such as puncture with a hypodermic needle) with strong muscular movements, and when exposed to a noxious environment (such as an acidic solution) show abnormal swimming behavior, attempts to jump from the water, and more rapid opercular movements. Such effects indicate some, perhaps considerable, distress, but it is not possible to state unequivocally that it is pain-induced distress.

Recent research has identified nociceptors in fish (Ashley et al. 2006, 2007; Sneddon 2002; Sneddon et al. 2003a) that are physiologically similar to mammalian nociceptors. In vivo administration of a noxious stimulus resulted in aberrant behaviors (rocking on the substrate and rubbing of the affected area) and adverse changes in physiology in rainbow trout over a period of 3 to 6 hours (Sneddon et al. 2003a,b); morphine reduced the incidence of these behaviors (Sneddon 2003; Sneddon et al. 2003b). Recent research has also shown that, after a one-time subcutaneous injection of 1% acetic acid to the lower and upper frontal lip, trout do not show appropriate neophobic or antipredator behaviors when compared to behavioral impairments associated with pain (Ashley et al. 2009; Sneddon et al. 2003b). Goldfish given electric shock display agitated swimming behavior but the threshold for this response increases if morphine is injected, while naloxone blocks the morphine effect (Jansen and Greene 1970). Work by Ehrensing and colleagues (1982) showed that the endogenous opioid antagonist MIF-1 downregulates sensitivity to opioids in goldfish, which then do not show an escape response to electric shock.

Studies have shown that goldfish are able to learn to avoid noxious, potentially painful stimuli such as electric shock (Portavella et al. 2002, 2004). Learned avoidance of a stimulus associated with a noxious experience has also been observed in other fish species including common carp and pike (*Esox lucius*), which avoided hooks in angling trials (Beukema 1970a,b; Overmier and Hollis 1983, 1990).

Amphibians

Amphibian species such as anurans (frogs and toads) and urodeles (salamanders) are commonly used in laboratory animal research settings (Schaeffer 1997), but there is no objective means to assess the presence and severity of pain in amphibians, especially since they do not exhibit any facial expression (Hadfield and Whitaker 2005). Some exotic animal clinicians use nonspecific clinical signs such as decrease in avoidance movement (e.g., when approached by a handler) or decrease in appetite as indicators of pain in these animals. Research has shown that amphibians are able and motivated to learn to avoid noxious stimuli (Strickler-Shaw and Taylor 1991).

CONCLUSIONS AND RECOMMENDATIONS

Further studies to develop robust, reliable, broadly applicable pain assessment tools are required. The general assumption is that the magnitude of the clinical signs and behavioral changes observed correlates closely with the intensity of pain. The extent to which these behavior-based assessments reflect the affective component of pain is uncertain and requires an improved understanding of the nature of pain, consciousness, and affective state in animals (see Box 1-2 in Chapter 1). Further, the lack of overlap between the assessment techniques used by veterinarians, pain researchers (Appendix A), and psychologists (Box 1-4) is an impediment to progress toward a broadly shared understanding.

The committee offers the following conclusions and recommendations:

1. Pain in animals is difficult to assess and greatly depends on the combination of a structured clinical examination and good knowledge of the normal appearance and behavior of the animals involved.
2. Observing animals' response to analgesic treatment can help refine clinical assessment schemes.
3. As more objective pain assessment schemes are developed, they should be adopted. The paucity of information for species other than farm animals, rats, and mice is detrimental to the animals' welfare and well-being as well as the quality of scientific research.
4. Responses of animals in analgesic drug tests and in models of pain can be used in efforts to identify (1) specific behaviors for use in assessment schemes and (2) sources of variation and factors that may influence pain intensity and analgesic efficacy.

REFERENCES

Arras M, Rettich A, Cinelli P, Kasermann HP, Burki K. 2007. Assessment of post-laparotomy pain in laboratory mice by telemetric recording of heart rate and heart rate variability. BMC Vet Res Aug 2:3-16.

Ashley FH, Waterman-Pearson AE, Whay HR. 2005. Behavioural assessment of pain in horses and donkeys: Application to clinical practice and future studies. Equine Vet J 37(6): 565-575.

Ashley PJ, Sneddon LU, McCrohan CR. 2006. Properties of corneal receptors in a teleost fish. Neurosci Lett 410(3):165-168.

Ashley PJ, Sneddon LU, McCrohan CR. 2007. Nociception in fish: Stimulus-response properties of receptors on the head of trout *Oncorhynchus mykiss*. Brain Res 1166:47-54.

Ashley PJ, Ringrose S, Edwards KL, Wallington E, McCrohan CR, Sneddon LU. 2009. Effect of noxious stimulation upon anti-predator responses and dominance status in rainbow trout. Anim Behav 77(2):403-410.

Beukema JJ. 1970a. Acquired hook-avoidance in the pike *Esox lucius L.* fished with artificial and natural baits. J Fish Biol 2:155-160.

Beukema JJ. 1970b. Angling experiments with carp: Decreased catchability through one trial learning. Neth J Zool 20:81-92.

Beynen AC, Baumans V, Bertens AP, Havenaar R, Hesp AP, Van Zutphen LF. 1987. Assessment of discomfort in gallstone-bearing mice: A practical example of the problems encountered in an attempt to recognize discomfort in laboratory animals. Lab Anim 21(1):35-42.

Brodbelt DC, Taylor PM, Stanway GW. 1997. A comparison of preoperative morphine and buprenorphine for postoperative analgesia for arthrotomy in dogs. J Vet Pharmacol Ther 20(4):284-289.

Cambridge AJ, Tobias KM, Newberry RC, Sarkar DK. 2000. Subjective and objective measurements of postoperative pain in cats. JAVMA 217(5):685-690.
Clark MD, Krugner-Higby L, Smith LJ, Heath TD, Clark KL, Olson D. 2004. Evaluation of liposome-encapsulated oxymorphone hydrochloride in mice after splenectomy. Comp Med 54(5):558-563.
Coggeshall RE. 2005. Fos, nociception and the dorsal horn. Prog Neurobiol 77(5):299-352.
Cowan A, Doxey JC, Harry EJ. 1977. The animal pharmacology of buprenorphine, an oripavine analgesic agent. Br J Pharmacol 60(4):547-554.
Dixon MJ, Robertson SA, Taylor PM. 2002. A thermal threshold testing device for evaluation of analgesics in cats. Res Vet Sci 72:205-210.
Driessen B, Zarucco L. 2007. Pain: From diagnosis to effective treatment. Clin Tech Equine P 6(2):126-134.
Duncan IJ, Beatty ER, Hocking PM, Duff SR. 1991. Assessment of pain associated with degenerative hip disorders in adult male turkeys. Res Vet Sci 50(2):200-203.
Ehrensing RH, Michell GF, Kastin AJ. 1982. Similar antagonism of morphine analgesia by MIF-1 and naloxone in *Carassius auratus*. Pharmacol Biochem Behav 17(4):757-761.
Firth AM, Haldane SL. 1999. Development of a scale to evaluate postoperative pain in dogs. JAVMA 214(5):651-659.
Flecknell PA. 1984. Relief of pain in laboratory animals. Lab Anim 18:147-160.
Flecknell PA, Liles JH. 1991. The effects of surgical procedures, halothane anaesthesia and nalbuphine on locomotor activity and food and water consumption in rats. Lab Anim 25(1):50-60.
Gabriel AF, Marcus MA, Honig WM, Walenkamp GH, Joosten EA. 2007. The CatWalk method: A detailed analysis of behavioral changes after acute inflammatory pain in the rat. J Neurosci Methods 163(1):9-16.
Gaynor JS, Muir W. 2002. Handbook of Veterinary Pain Management. St. Louis, MO: Mosby.
Giamberardino MA, Valente R, de Bigontina P, Vecchiet L. 1995. Artificial ureteral calculosis in rats: Behavioural characterization of visceral pain episodes and their relationship with referred lumbar muscle hyperalgesia. Pain 61(3):459-469.
Gonzalez MI, Field MJ, Bramwell S, McCleary S, Singh L. 2000. Ovariohysterectomy in the rat: A model of surgical pain for evaluation of pre-emptive analgesia? Pain 88(1):79-88.
Grisneaux E, Pibarot P, Dupuis J, Blais D. 1999. Comparison of ketoprofen and carprofen administered prior to orthopedic surgery for control of postoperative pain in dogs. JAVMA 215(8):1105-1110.
Hadfield CA, Whitaker BR. 2005. Amphibian emergency medicine and care. Semin Avian Exot Pet 14(2 Spec Iss):79-89.
Harvey-Clark CJ, Gilespie K, Riggs KW. 2000. Transdermal fentanyl compared with parenteral buprenorphine in post-surgical pain in swine: A case study. Lab Anim 34(4):386-398.
Hawkins P. 2002. Recognizing and assessing pain, suffering and distress in laboratory animals: A survey of current practice in the UK with recommendations. Lab Anim 36(4):378-395.
Hay M, Vulin A, Génin S, Sales P, Prunier A. 2003. Assessment of pain induced by castration in piglets: Behavioral and physiological responses over the subsequent 5 days. Appl Anim Behav Sci 82(3):201-218.
Hayes KE, Raucci JA Jr, Gades NM, Toth LA. 2000. An evaluation of analgesic regimens for abdominal surgery in mice. Contemp Top Lab Anim 39(6):18-23.
Hazewinkel HAW, van den Brom WE, Theyse LFH, Pollmeier M, Hanson PD. 2008. Comparison of the effects of firocoxib, carprofen and vedaprofen in a sodium urate crystal induced synovitis model of arthritis in dogs. Res Vet Sci 84(1):74-79.
Hernandez J, Shearer JK, Webb DW. 2002. Effect of lameness on milk yield in dairy cows. JAVMA 220(5):640-644.

Hernandez JA, Garbarino EJ, Shearer JK, Risco CA, Thatcher WW. 2005. Comparison of milk yield in dairy cows with different degrees of lameness. JAVMA 227(8):1292-1296.

Hess A, Sergejeva M, Budinsky L, Zeilhofer HU, Brune K. 2007. Imaging of hyperalgesia in rats by functional MRI. Eur J Pain 11(1):109-119.

Hogan Q. 2002. Animal pain models. Reg Anesth Pain Med 27(4):385-401.

Holton L, Reid J, Scott EM, Pawson P, Nolan A. 2001. Development of a behaviour-based scale to measure acute pain in dogs. Vet Rec 148(17):525-531.

Holton LL, Scott EM, Nolan AM, Reid J, Welsh E. 1998. Relationship between physiological factors and clinical pain in dogs scored using a numerical rating scale. J Small Anim Pract 39(10):469-474.

Hudson C, Whay H, Hixley J. 2008. Recognition and management of pain in cattle. In Practice 30:126-134.

ILAR. 2009. Pain and Distress in Fish. ILAR J 50:327-418.

Jansen GA, Greene NM. 1970. Morphine metabolism and morphine tolerance in goldfish. Anesthesiology 32(3):231-235.

Karas AZ. 2002. Postoperative analgesia in the laboratory mouse, *Mus musculus*. Lab Anim (NY) 31(7):49-52.

KuKanich B, Lascelles BD, Papich MG. 2005. Assessment of a von Frey device for evaluation of the antinociceptive effects of morphine and its application in pharmacodynamic modeling of morphine in dogs. Am J Vet Res 66:1616-1622.

Lascelles BD, Capner C, Waterman AF. 1995. Survey of perioperative analgesic use in small animals. Vet Rec 137(26):676.

Lascelles BD, Cripps PJ, Jones A, Waterman AE. 1997. Post-operative central hypersensitivity and pain: The pre-emptive value of pethidine for ovariohysterectomy. Pain 73(3):461-471.

Le Bars D, Gozariu M, Cadden SW. 2001. Animal models of nociception. Pharmacol Rev 53(4):597-652.

Leach MC, Allweiler S, Richardson C, Roughan JV, Flecknell PA. 2009. Behavioural effects of ovarohysterectomy and oral administration of meloxicam in laboratory housed rabbits. Res Vet Sci 87(2):336-347.

Lester SJ, Mellor DJ, Ward RN, Holmes RJ. 1991. Cortisol responses of young lambs to castration and tailing using different methods. New Zeal Vet J 39(4):134-138.

Lester SJ, Mellor DJ, Holmes RJ, Ward RN, Stafford KJ. 1996. Behavioural and cortisol responses of lambs to castration and tailing using different methods. New Zeal Vet J 44:45-54.

Ley SJ, Waterman AE, Livingston A. 1996. Measurement of mechanical thresholds, plasma cortisol and catecholamines in control and lame cattle: A preliminary study. Res Vet Sci 61:172-173.

Liles JH, Flecknell PA. 1992. The effects of buprenorphine, nalbuphine and butorphanol alone or following halothane anaesthesia on food and water consumption and locomotor movement in rats. Lab Anim 26(3):180-189.

Liles JH, Flecknell PA. 1993a. The effects of surgical stimulus on the rat and the influence of analgesic treatment. Br Vet J 149(6):515-525.

Liles JH, Flecknell PA. 1993b. The influence of buprenorphine or bupivacaine on the post-operative effects of laparotomy and bile-duct ligation in rats. Lab Anim 27(4):374-380.

Lykkegaard K, Lauritzen B, Tessem L, Weikop P, Svendsen O. 2005. Local anaesthetics attenuates spinal nociception and HPA-axis activation during experimental laparotomy in pigs. Res Vet Sci 79(3):245-251.

Malavasi LM, Nyman G, Augustsson H, Jacobson M, Jensen-Waern M. 2006. Effects of epidural morphine and transdermal fentanyl analgesia on physiology and behaviour after abdominal surgery in pigs. Lab Anim 40(1):16-27.

Mathews KA, Pettifer G, Foster R, McDonell W. 2001. Safety and efficacy of preoperative administration of meloxicam, compared with that of ketoprofen and butorphanol in dogs undergoing abdominal surgery. Am J Vet Res 62(6):882-888.

Mellor DJ, Stafford KJ. 2000. Acute castration and/or tailing distress and its alleviation in lambs. New Zeal Vet J 48(2):33-43.

Mogil JS, Wilson SG, Bon K, Lee SE, Chung K, Raber P, Pieper JO, Hain HS, Belknap JK, Hubert L, Elmer GI, Chung JM, Devor M. 1999. Heritability of nociception I: Responses of 11 inbred mouse strains on 12 measures of nociception. Pain 80(1-2):67-82.

Molony V, Kent JE, McKendrick IJ. 2002. Validation of a method for assessment of an acute pain in lambs. Appl Anim Behav Sci 76(3):215-238.

Molony V, Kent JE, Robertson IS. 1995. Assessment of acute and chronic pain after different methods of castration of calves. Appl Anim Behav Sci 46:33-48.

Morgan D, Cook CD, Picker MJ. 1999. Sensitivity to the discriminative stimulus and antinociceptive effects of mu opioids: Role of strain of rat, stimulus intensity, and intrinsic efficacy at the mu opioid receptor. J Pharmacol Exp Ther 289(2):965-975.

Morton DB, Griffiths PH. 1985. Guidelines on the recognition of pain, distress and discomfort in experimental animals and an hypothesis for assessment. Vet Rec 116(16):431-436.

Morton CM, Reid J, Scott EM, Holton LL, Nolan AM. 2005. Application of a scaling model to establish and validate an interval level pain scale for assessment of acute pain in dogs. Am J Vet Res 66(12):2154-2166.

Noonan GJ, Rand JS, Priest J, Ainscow J, Blackshaw JK. 1994. Behavioral observation of piglets undergoing tail docking, teeth clipping and ear notching. Appl Anim Behav Sci 39:203-213.

NRC (National Research Council). 1992. Recognition and Alleviation of Pain and Distress in Laboratory Animals. Washington: National Academy Press.

NRC. 2008. Recognition and Alleviation of Distress in Laboratory Animals. Washington: National Academies Press.

Overmier JB, Hollis KL. 1983. Teleostean telencephalon in learning. In: David RE, Northcutt G, eds. Fish Neurobiology Vol II: Higher Brain Areas and Functions. Ann Arbor: University of Michigan. pp. 265-284.

Overmier JB, Hollis KL. 1990. Fish in the think tank: Learning, memory and integrated behavior. In: Olton D, Kesner R, eds. The Neurobiology of Comparative Cognition. Hillsdale, NJ: Erlbaum. pp. 205-236.

Parga, ML. 2002. Assessment of the efficacy of meloxicam and development of a pain-scoring system based on behaviour in rabbits undergoing elective surgery. In: Proceedings of the 45th Annual BSAVA Congress Congress, 4th-7th April 2002, Birmingham, UK.

Portavella M, Vargas JP, Torres B, Salas C. 2002. The effects of telencephalic pallial lesions on spatial, temporal, and emotional learning in goldfish. Brain Res Bull 57(3-4):397-399.

Portavella M, Torres B, Salas C. 2004. Avoidance response in goldfish: Emotional and temporal involvement of medical and lateral telencephalic pallium. J Neurosci 24(9):2335-2342.

Price J, Catriona S, Welsh EM, Waran NK. 2003. Preliminary evaluation of a behaviour-based system for assessment of post-operative pain in horses following arthroscopic surgery. Vet Anaesth Analg 30(3):124-137.

Puppe B, Schön PC, Tuchscherer A, Manteuffel G. 2005. Castration-induced vocalisation in domestic piglets, *Sus scrofa*: Complex and specific alterations of the vocal quality. Appl Anim Behav Sci 95(1-2):67-78.

Pypendop BH, Ilkiew JE, Robertson SA. 2006. Effect of intravenous administration of lidocaine on the thermal threshold in cats. Am J Vet Res 67:16-20.

Ranger M, Johnston CC, Anand KJS. 2007. Current controversies regarding pain assessment in neonates. Sem Perinatol 31(5):283-288.

Reilly SC, Quinn JP, Cossins AR, Sneddon LU. 2008. Behavioural analysis of a nociceptive event in fish: Comparisons between three species demonstrate specific responses. Appl Anim Behav Sci 114:248-249.

Rietmann TR, Stauffacher M, Bernasconi P, Auer JA, Weishaupt MA. 2004. The association between heart rate, heart rate variability, endocrine and behavioural pain measures in horses suffering from laminitis. J Vet Med A Physiol Pathol Clin Med 51(5):218-225.

Rose JD. 2002. The neurobehavioral nature of fishes and the question of awareness and pain. Rev Fish Sci 10:1-38.

Roughan JV, Flecknell PA. 2000. Effects of surgery and analgesic administration on spontaneous behaviour in singly housed rats. Res Vet Sci 69(3):283-288.

Roughan JV, Flecknell PA. 2001. Behavioural effects of laparotomy and analgesic effects of ketoprofen and carprofen in rats. Pain 90(1-2):65-74.

Roughan JV, Flecknell PA. 2002. Buprenorphine: A reappraisal of its antinociceptive effects and therapeutic use in alleviating post-operative pain in animals. Lab Anim 36(3):322-343.

Roughan JV, Flecknell PA. 2003. Evaluation of a short duration behaviour-based post-operative pain scoring system in rats. Eur J Pain 7(5):397-406.

Roughan JV, Flecknell PA. 2004. Behaviour-based assessment of the duration of laparotomy-induced abdominal pain and the analgesic effects of carprofen and buprenorphine in rats. Behav Pharmacol 15(7):461-472.

Roughan JV, Flecknell PA. 2006. Training in behaviour-based post-operative pain scoring in rats—An evaluation based on improved recognition of analgesic requirements. Appl Anim Behav Sci 96(3-4):327-342.

Schaeffer D. 1997. Anesthesia and analgesia. In: Kohn DF, ed. Nontraditional Laboratory Animal Species in Anesthesia and Analgesia in Laboratory Animals. San Diego: Academic Press.

Sharp J, Zammit T, Azar T, Lawson D. 2003. Recovery of male rats from major abdominal surgery after treatment with various analgesics. Contemp Top Lab Anim Sci. 42(6):22-27.

Slingsby LS, Waterman-Pearson AE. 1998. Comparison of pethidine, buprenorphine and ketoprofen for postoperative analgesia after ovariohysterectomy in the cat. Vet Rec 143(7):185-189.

Slingsby LS, Jones A, Waterman-Pearson AE. 2001. Use of a fingermounted device to compare mechanical nociceptive thresholds in cats given pethidine or no medication after castration. Res Vet Sci 70:243-246.

Sneddon LU. 2002. Anatomical and electrophysiological analysis of the trigeminal nerve in a teleost fish, *Oncorhynchus mykiss*. Neurosci Lett 319(3):167-171.

Sneddon LU. 2003. The evidence for pain perception in fish: The use of morphine as an analgesic. Appl Anim Behav Sci 83:153-162.

Sneddon LU. 2006. Ethics and welfare: Pain perception in fish. B Eur Assoc Fish Pat 26(1): 6-10.

Sneddon LU, Braithwaite VA, Gentle MJ. 2003a. Do fishes have nociceptors? Evidence for the evolution of a vertebrate sensory system. Proc R Soc Lond B Biol Sci 270(1520): 1115-1121.

Sneddon LU, Braithwaite VA, Gentle MJ. 2003b. Novel object test: Examining nociception and fear in the rainbow trout. J Pain 4(8):431-440.

Stafford KJ, Mellor DJ. 2005. Dehorning and disbudding distress and its alleviation in calves. Vet J 169(3):337-349.

Strickler-Shaw S, Taylor DH. 1991. Lead inhibits acquisition and retention learning in bullfrog tadpoles. Neurotoxicol Teratol 13:167-173.

Svendsen O, Kok L, Lauritzan B. 2007. Nociception after intraperitoneal injection of a socium pentobarbitone formulation with and without lidocain in rats quantified by expression of neuronal c-fos in the spinal cord: A preliminary study. Lab Anim 41(2):197-203.

Veissier II, Rushen J, Colwell D, de Passillé AM. 2000. A laser-based method for measuring thermal nociception of cattle. Appl Anim Behav Sci 66(4):289-304.

Viñuela-Fernández I, Jones E, Welsh EM, Fleetwood-Walker SM. 2007. Pain mechanisms and their implication for the management of pain in farm and companion animals. Vet J 174(2):227-239.

Weary DM, Braithwaite LA, Fraser D. 1998. Vocal response to pain in piglets. Appl Anim Behav Sci 56(2-4):161-172.

Welsh EM, Nolan AM. 1995. Effect of flumixin meglumine on the threshold to mechanical stimulation in healthy and lame sheep. Res Vet Sci 58:61-66.

Whiteside GT, Harrison J, Boulet J, Mark L, Pearson M, Gottshall S, Walker K. 2004. Pharmacological characterisation of a rat model of incisional pain. Br J Pharmacol 141(1):85-91.

Wright-Williams SL, Courade JP, Richardson CA, Roughan JV, Flecknell PA. 2007. Effects of vasectomy surgery and meloxicam treatment on faecal corticosterone levels and behaviour in two strains of laboratory mouse. Pain 130(1-2):108-118.

Zwakhalen SMG, Hamers JPH, Berger MPF. 2006. The psychometric quality and clinical usefulness of three pain assessment tools for elderly people with dementia. Pain 126(1-3): 210-220.

4

Effective Pain Management

This chapter presents an overview of the basic clinical strategies, both pharmacologic and nonpharmacologic, for managing pain in laboratory animals. Topics include preventive analgesia, consequences of unrelieved pain, and ethical considerations relating to pain as a subject of study. Available information on pain management of nonmammalian species is also presented.

INTRODUCTION

The regulatory review process (see Appendix B) requires that investigators adequately control pain in research animals, unless procedures that may cause more than momentary or slight pain are justified for scientific reasons and approved by the IACUC. In order to treat or prevent pain, it is necessary to evaluate its source and intensity (for additional discussion see Chapter 3). As a rule, pain is likely to occur in proportional terms as a result of tissue injury—more extensive tissue damage results in greater pain and thus a need for a stronger analgesic regimen. While certain conditions reliably cause severe pain (e.g., acute nerve compression, burns, spastic contraction of smooth muscle) and inflammation often contributes to the worsening of pain, scientists do not fully understand how much pain to expect in various animal species. Information about the cause and effect of surgery or disease and pain in clinical veterinary medicine is largely based on observation and anecdote and tends to focus on commonly treated species, such as dogs, cats, and horses. Table 1-1 of Chapter 1 lists examples of typically painful conditions that occur either spontaneously or as a result of experimental procedure.

CLINICAL VETERINARY PAIN MANAGEMENT

The principles of clinical veterinary pain management and prevention, summarized in Boxes 4-1 and 4-2, are comparatively easy to apply in clinically familiar species such as dogs and cats, for which ranges of doses and drug combinations are relatively well known. However, the application of the principles discussed below to other laboratory animal species is a matter of trial and error until adequate scientific information is available to establish evidence-based guidelines, including information on the feasibility of various routes of administration (e.g., oral bioavailability, palatability, transdermal preparations). Readers are encouraged to seek publications (including the American College of Veterinary Anesthesiologists' Position Paper on the Treatment of Pain [ACVA 1998]), reports, books, and the veterinary literature for specific information on available drugs, doses, routes of administration, side effects, contraindications, and the like that may be useful for dogs, cats, rabbits, and other species used as research animals.

BOX 4-1
Current Guidelines for Clinical Veterinary Pain Management

- Sedation does not provide pain relief and may mask the animal's response to pain.
- Use of analgesic and adjunct drugs should be at effective plasma/tissue concentrations especially when the nociceptive barrage and pain are greatest (i.e., after surgery or injury).
- Use of more than one type of management strategy (e.g., multimodal analgesia [targeting multiple pain mechanisms with the use of local anesthetics and opioids] or anxiolytics when postsurgical pain is likely to be moderate to severe) is recommended.
- Avoidance of peaks and valleys in analgesic dosing (best accomplished by the administration of continuous or overlapping regimes) when postsurgical pain is expected to be severe maintains animal well-being.
- Monitoring, at appropriate intervals, of the effectiveness of analgesics administered is crucial.
- If there is doubt about the source of an animal's clinical signs, administration of an additional dose of analgesic—dependent on the drug, species, and often the individual animal—can help determine whether pain was the cause (while this is not commonly done in laboratory animal medicine, this method of pain control/alleviation in nonrodent species is common in clinical veterinary practice in a patient-specific manner).

BOX 4-2
Additional Considerations for the Prevention and Management of Pain in Laboratory Animals

- Pain in animals is often unrecognized and undertreated.
- If a procedure is considered painful in humans, it should be assumed to be painful in laboratory animals, regardless of their age or species.
- Adequate treatment of pain may be associated with decreased complications, lower mortality, reduced variability in experimental data, and improved scientific outcomes.
- The appropriate use of environmental, nonpharmacologic, or pharmacological interventions, as well as the selection of humane endpoints in animal experimentation, can prevent or reduce animal pain in most experimental designs without compromising the scientific validity of the research, except where pain is the subject of research.
- Researchers, veterinarians, and animal care professionals should be responsible for learning about the assessment, prevention, and management of pain in laboratory animals.
- Veterinarians and animal care professionals should develop IACUC-approved educational guidelines and protocols for the management of pain in laboratory animals at their institution.

Some ranges for effective doses of analgesics in rats and mice (i.e., doses that reduce experimental measures of pain and/or reach tissue concentrations believed to be effective in other species) are available through literature search. However, strain differences in animals' responses to analgesics and anesthetics are an important factor to consider (Mogil et al. 2005; Terner et al. 2003; Wilson et al. 2003a,b).

STRATEGIES FOR MANAGING PAIN IN LABORATORY ANIMALS

Effective management of pain in laboratory animals often begins with general (surgical) anesthesia, but also includes local anesthetics, analgesics, anxiolytics, and sedatives as well as nonpharmacological methods (including minimization of tissue trauma). Pain management goals range from total elimination (as, for example, during general anesthesia for a surgical procedure) to pain that is tolerated without compromising the animal's well-being.

General Anesthesia

When animals are anesthetized for procedures that would otherwise cause pain, it is important to maintain an appropriate depth of anesthesia.

A wide range of indices have been developed to assess depth of anesthesia in animals and humans (Appadu and Vaidya 2008; Bruhn et al. 2006; Franks 2008; John and Prichep 2005; Lu et al. 2003; Murrell and Johnson 2006; Otto 2008; Whelan and Flecknell 1992); these include autonomic responses such as changes in heart rate and blood pressure, alterations in the EEG or other measures of CNS function, or changes in somatic reflex responses to noxious stimuli. During anesthesia not accompanied by neuromuscular blocking agents, depression of somatic reflex responses is the most widely used method for ensuring an appropriate depth of anesthesia. In all animal species, absence of the pedal withdrawal reflex indicates a surgical plane of anesthesia (i.e., anesthesia that is deep enough to eliminate the experience of pain and thus allow surgery to take place). Although this is an easily assessed index, it is important to use a stimulus that is sufficiently noxious but not so strong as to produce tissue damage. In some species, other reflexes, such as the response to applying a clamp to the nasal septum (pigs) or pinching the ears (rabbit, guinea pig), are also useful but reliance on these responses has been criticized (Antognini et al. 2005) because animals may lose consciousness at much lighter anesthesia planes, in which case the persistence of reflexes would not indicate pain perception (see also Box 1-3 in Chapter 1). Doses of anesthetic agents sufficient to suppress spinal reflexes may therefore be greater than those required to carry out surgery humanely; if these reflexes are not suppressed, surgery will be hampered by the animals' repeated reflex movements. Although the use of neuromuscular blocking agents (which prevent neurotransmitters from acting on their receptors in skeletal muscles) could prevent such movements, it would also require intubation and mechanical ventilation of the animal. For practical reasons, suppression of withdrawal responses remains the most useful means of ensuring loss of both awareness and responses to surgical stimuli.

The ideal general anesthetic should rapidly and/or smoothly induce muscle relaxation and a surgical plane of anesthesia, and should be readily controllable and reversible. There are two categories of general anesthetics used in laboratory animal medicine: volatile inhalants (e.g., isoflurane) and injectable drugs (e.g., barbiturates, other sedative-hypnotic agents such as propofol, or combinations of drugs such as propofol-fentanyl). The latter category also includes total intravenous anesthesia (TIVA). TIVA techniques may be useful in laboratory animal settings where the equipment required for inhalant anesthesia is not practical or possible (e.g., near MRI units). Other injectable general anesthetic drugs still in use due to their unique application in specialized studies include α-chloralose, tribromoethanol, and urethane. These drugs have certain specific applications but may not be appropriate for situations in which animals will recover (Gaertner et al. 2008; Karas and Silverman 2006; Koblin 2002; Meyer and Fish 2005) as, after surgery, with anesthetic withdrawal and recovery, the animals will experience pain unless they receive analgesics.

Sedation/Anxiolysis

Sedatives and anxiolytics are adjuncts to general anesthetics and are also used in pain management strategies. These two distinct classes of drugs are often used in combination to modulate, block, or relieve pain. Terminology varies but a general distinction between the sedative-hypnotic agents and anxiolytics is often useful. Sedative-hypnotic drugs (e.g., barbiturates and drugs with significant sedating properties such as α_2-adrenoreceptor agonists) produce dose-dependent states of CNS depression that vary from somnolence to general anesthesia and even death. Anxiolytics are drugs that reduce anxiety or fear (e.g., benzodiazepines) and can induce sleep. Some anxiolytic drugs, previously termed "tranquilizers" (e.g., phenothiazines like acepromazine and butyrophenones like haloperidol and droperidol), produce a state of relaxation and indifference to external stimuli and, in elevated doses, can induce an undesirable cataleptic state rather than general anesthesia. Of the above drugs and classes, only the α_2-adrenoreceptor agonists have analgesic efficacy. Neither barbiturates nor anxiolytics are analgesic; barbiturates may in fact contribute to a hyperalgesic state, while phenothiazines and butyrophenones are generally considered devoid of analgesic efficacy. Readers are referred to the section "Modulatory Influences on Pain: Anxiety, Fear, and Stress" in Chapter 2 for a discussion of the relationship of anxiety and pain.

Neuroleptanalgesia is an intense analgesic and amnesic state produced by the combination of an opioid analgesic and a neuroleptic drug (this description is adapted from the *American Heritage Medical Dictionary* 2007). The neuroleptic drug component is a phenothiazine or butyrophenone (or possibly an anxiolytic) and the analgesic is a potent and efficacious opioid that also acts as a major tranquilizer (i.e., anxiolytic). Butorphanol-acepromazine, fentanyl-fluanisone (Hypnorm®[1]), and oxymorphone-midazolam are examples of commonly used veterinary neuroleptanalgesic combinations. Neuroleptanalgesic combinations by themselves are not sufficient for most surgical interventions. However, the use of drugs with sedative or tranquilizing properties (neurolepts as well as α_2-adrenoreceptor agonists) combined with opioids, ketamine, or tiletamine-zolazepam (Telazol®) can cause states ranging from modified consciousness (e.g., reduction of anxiety or "conscious sedation") to complete unconsciousness (general anesthesia). Table 4-1 summarizes the analgesic properties of selected drugs, includ-

[1]Hypnorm is not available in the United States (as of August 2009).

TABLE 4-1 Analgesic Properties of Selected Anesthetic Drugs and Adjuncts

Drug	Class	Analgesic Efficacy
α_2-Adrenoreceptor agonists	Analgesic/sedative-hypnotic	Yes
Barbiturates	Sedative-hypnotic	No
Benzodiazepines	Anxiolytic	No
Butyrophenones	Neuroleptic/anxiolytic	No
Chloralose, chloral hydrate	Sedative-hypnotic	No
Halogenated inhalant anesthetics	General anesthetic	No
Ketamine	Dissociative, NMDA antagonist	Yes
Nitrous oxide	General anesthetic (human); general anesthetic adjunct only in animals	Yes
Opioids	Analgesic	Yes
Phenothiazines	Neuroleptic/anxiolytic	No
Propofol	Sedative-hypnotic	No
Tiletamine-zolazepam (Telazol®)	Combination of a dissociative/ NMDA receptor antagonist and a benzodiazepine anxiolytic	Yes
Tribromoethanol	Sedative-hypnotic	No
Urethane (e.g., ethyl carbamate)	Not classified	No

NOTE: Drugs with inherent analgesic effects may contribute to postoperative pain control but are not sufficient to exert such control in and of themselves.

ing tranquilizers, sedatives, and anesthetics, commonly used in laboratory animals.

Analgesia

Conventional analgesic drug classes include opioids, NSAIDs, and local anesthetics. Although analgesia is defined as "lack of pain," complete elimination of pain in awake animals is commonly neither achievable nor desirable. Pain has a protective role as it usually serves to limit further injury; for example, humans with no skin sensation are prone to undetectable injury or infection. But in some instances animals with untreated severe pain may struggle or self-mutilate and exacerbate or cause additional injury to themselves. With most analgesic techniques, however, residual pain naturally limits activity, although it is not a restraint mechanism and should not be used to restrain animals.

The goal of analgesic drug intervention is to achieve a balanced state during which an animal is neither substantially hindered by pain nor adversely affected by the side effects of analgesics. Often the use of a single analgesic is sufficient. An emerging practice for the prevention or treatment of established pain in both human and veterinary patients, however, is the combined use of two or more types of analgesics, or "multimodal analgesia" (Buvanendran and Kroin 2007; Corletto 2007; Hellyer et al. 2007; Kehlet et al. 2006; Lemke 2004; White 2005; White et al. 2007). Multimodal post-surgical analgesia may be regarded as overly complicated, but cited benefits include more effective and efficient analgesia and possible dose reduction of one or more individual drugs.

In theory, treatment of patients with nonopioid analgesics to reduce the overall requirement for opioids would result in fewer opioid-induced side effects. The concept, known as "opioid sparing," is a desirable goal because extended or high-dose opioid therapy is often accompanied by unwanted side effects (e.g., sedation, constipation, urinary retention, or analgesic tolerance) that prolong or complicate convalescence (Kehlet 2004; White et al. 2007). Synergy (i.e., greater analgesia than predicted from a simple additive effect of the combination of two drugs acting with different mechanisms) has been demonstrated in numerous experimental animal models (e.g., Price et al. 1996; Kolesnikov et al. 2000; Matthews and Dickenson 2002; Qiu et al. 2007) as well as with combinations of opioids, NSAIDs, local anesthetics, α_2-agonists, ketamine, tramadol, and gabapentin (Guillou et al. 2003; Koppert et al. 2004; Reuben and Buvanendran 2007; White et al. 2007). Multimodal analgesia using "adjuvant analgesics" (i.e., antidepressants, antiepileptic drugs, NMDA antagonists, or transdermal lidocaine) may also be an effective alternative for the treatment of refractory chronic pain unresponsive to the administration of a single agent (Knotkova and Pappagallo 2007). Table 4-2 summarizes pharmacologic methods for treating pain of various intensities.

Advanced Analgesic Techniques

The ability to provide analgesia to laboratory animals is limited by the lack of information about species-specific drug effects and doses. It is perhaps useful to understand the state-of-the-art techniques currently used in clinical (i.e., nonlaboratory) veterinary medicine as a potential objective for laboratory animal pain medicine; identification of the most useful techniques may lead to important innovations to help overcome barriers to the provision of analgesia. Needless to say, size, species, and technical aspects will continue to be limiting factors for many techniques. Box 4-3 provides a summary of analgesic techniques and their limitations.

TABLE 4-2 Pharmacologic Approach to Pain Management Based on Predicted Intensity

Pain Intensity	Analgesic Approach
Low	Single-agent therapy acceptable NSAIDs, local anesthetic infiltration, or opioid agonist-antagonists (butorphanol, buprenorphine)
Moderate	Multimodal analgesia to be considered NSAIDs in combination with adjuncts such as local anesthetics, opioid agonist-antagonists (buprenorphine), tramadol, α_2-agonists, NMDA antagonists
High	Multimodal analgesia recommended mu-opioid agonists (morphine, hydromorphone, fentanyl, methadone) + one or more of the following: NSAIDs, local anesthetics, α_2-agonists, antiepileptic drugs, NMDA antagonists Advanced analgesic techniques: epidural administration of local anesthetics with or without opioids and constant rate infusions

Nonpharmacologic Methods

Nonpharmacologic approaches to pain management are appropriate when the use of pharmacological methods is contraindicated, when effective analgesic drugs are not available, or to complement drug therapy. Nonpharmacologic methods include preventive strategies that help minimize causative factors for pain, through, for example, appropriate animal handling and minimization of tissue trauma during surgery. Such techniques are important because both long-duration surgery and extensive tissue manipulation (e.g., rib retraction, prolonged tourniquet-induced limb ischemia, disproportionately long incision relative to animal size) result in increased postoperative pain. Training in proper surgical techniques coupled with knowledge of comparative anatomy is necessary to appreciate the distinct needs of each animal species before, during, and after surgery and to uphold the 3Rs principle of refinement. Moreover, nonphysiologic restraint or surgical positioning of animals may exert undue pressure on joints, nerves, or soft tissues and cause significant postprocedural pain. These sources of pain are avoidable if investigators and animal care personnel are trained to understand that *any* form of tissue pressure, damage, or ischemia is a potential cause of pain (Martini et al. 2000; LASA 1990). Minimally invasive surgery techniques (e.g., fiberoptic technologies) reduce tissue injury and are associated with reduced postsurgical pain, stress response, and convalescence time compared to open or scalpel surgery (reviewed by Karas et al. 2008).

BOX 4-3
Advanced Analgesic Techniques

- Low-dose epidural administration of opioids or opioid-local anesthetic combinations can result in analgesia whose quality is similar to if not better than that achieved with systemic administration. This method depends on technical expertise and may be challenging to implement in very small animals. Epidural administration of drugs has not been studied in nonmammalian vertebrates.
- Local anesthetics can be injected into joints, wounds, and body cavities (abdominal or pleural) by continuous or intermittent injection through intra-wound catheters, greatly reducing the need for systemic administration of other analgesics (Liu et al. 2006). The relatively short duration of the action of local anesthetics may limit their utility in situations where redosing is difficult. Lidocaine is used intravenously to provide analgesia after tissue injury (Omote 2007).
- Oral administration of some analgesics is feasible (e.g., NSAIDs, opioids, gabapentin), but for some drugs (opioids) first-pass (species-dependent) metabolism limits bioavailability, necessitating dose adjustment, use of a different route of administration, or selection of another drug. Compounding of drugs into palatable forms that animals are willing to consume is possible, but without data to support a particular method, one must be concerned about absorption, shelf life, and efficacy.
- Dilution of injectable analgesics to make them easier to use or to improve provision in very small animals must be done with the understanding that formulations may not work as well and that shelf life is not predictable.
- Continuous infusion of certain types of analgesics (e.g., opioids, ketamine, α_2-adrenoreceptor agonists) avoids "peaks and valleys" in drug concentration and may provide better coverage for moderate to severe pain. Transdermal preparations are available in formulations suitable for larger animals and may be useful in producing uninterrupted analgesia. Sustained-release formulations make it possible to avoid periods of inadequate drug administration. For further consultation please see Carroll 2008; Flecknell 2009; Gaynor and Muir 2002; Hellyer et al. 2007; Krugner-Higby et al. 2008; Lamont and Mathews 2007; Robertson 2005; Tranquilli et al. 2007; Valverde and Gunkel 2005.

METHODS FOR THE PREVENTION OR MANAGEMENT OF PAIN

While classic pharmacologic treatment requires drugs with specific analgesic properties, unconventional drugs, such as antiepileptics, can also be effective. And when anxiety contributes to pain, drugs with anxiolytic properties can be added.

Analgesics

A thorough review of the effects and doses of analgesic drugs is beyond the scope of this work (for comprehensive reviews see Carroll 2008; Flecknell and Waterman-Pearson 2000; Gaynor and Muir 2002; Hawk et al. 2005; Lamont and Mathews 2007; Robertson 2005; Valverde and Gunkel 2005). Instead, this section provides an overview of analgesic drugs that are currently used or may become useful in laboratory animal medicine.

Opioids

Opioid analgesics are important drugs for surgical analgesia and/or therapeutic management of moderate to severe pain in humans and certain animal species. There are two general categories of such analgesics (Ross et al. 2006; Stefano et al. 2005; Waldhoer et al. 2004): opioid receptor agonists (e.g., morphine, hydromorphone, fentanyl) and mixed opioid receptor agonist/antagonists (e.g., buprenorphine, butorphanol); the latter group possesses (in a single molecule) agonist efficacy at one of the three types of opioid receptor and antagonist efficacy at a different opioid receptor.

A third group of endogenous opioid peptides (e.g., endorphins, enkephalins, and dynorphins) are produced by the body and also act on opioid receptors. It is a misconception, however, to assume that the only role of endogenous opioid peptides is to produce analgesia; they have multiple, nonanalgesic functions depending on where in the body they are produced and released. Given the existence of three distinct opioid receptors, all located in variable densities in various tissues, differences in the selectivity and affinity of opioid drugs and endogenous opioid peptides are believed to account for many of the variations in the effect profile of opioids (Fields 2004; Waldhoer et al. 2004). And because opioid receptors are subject to regulation (e.g., by phosphorylation or endocytosis), the effects of both endogenous and exogenous opioids can be influenced by the "state" of the receptor. Changes such as these presumably account for the phenomenon of analgesic tolerance, a reduction in the analgesic effectiveness of a given dose of drug after repeated administration.

Opioids are the most efficacious analgesics available, but their use is accompanied by undesirable effects that include an increase in smooth muscle tone and reduction in propulsive motility of the gastrointestinal tract (leading to constipation), cough suppression, respiratory depression, behavioral changes (euphoria and dysphoria, excitement, or increased locomotion), and physiological dependence. In addition to their presence on neurons both in the nociceptive pathway (see Chapter 2) and elsewhere in the body (e.g., the gastrointestinal tract), opioid receptors are found on cells of the immune system and opioid effects on immune function vary from stimulation to inhibition (Stefano et al. 2005; Page et al. 2001). In rats and other rodents, pica (the ingestion of nonedible substances, such as bedding) and the consumption of large volumes of food have been noted with the use of the partial opioid receptor agonist/weak antagonist buprenorphine (Aung et al. 2004; Bosgraaf et al. 2004; Clark et al. 1997; Yamamoto et al. 2004). Concern about the undesirable side effects of opioids is frequently cited as a reason for not using them, but for limited or short-term therapy the side effects are often either manageable or not a problem.

Dose regimens of opioid analgesics for dogs, cats, horses, rats, mice, a few species of birds, and sheep have been reported. When such regimens are based on experimental evidence, it frequently derives from an analgesiometric testing method (such as thermal threshold; Johnson et al. 2007; Robertson et al. 2005a,b; Waterman et al. 1991; Wilson et al. 2003a,b). Doses for other mammals currently listed in formularies are based on extrapolation. Relatively little is known about the efficacy, drug choices, or side effects of opioids in amphibians, reptiles, invertebrates, and most birds.

In addition to classical intravenous, intramuscular, and intraperitoneal routes of administration, many opioids are also substantially bioavailable by nasal, sublingual, or rectal routes (Lindhardt et al. 2000; Robertson et al. 2005a). Oral administration of opioids in mammals often diminishes their bioavailability, making this method of delivery less effective. Additionally, long-duration formulations of opioids have been investigated in animal models and, although not yet commercially available, may represent a future method to provide sustained analgesia in laboratory animals (Krugner-Higby et al. 2008; Smith et al. 2004).

Because of the relative safety of opioids, information about effective dose ranges and novel methods of administration would be useful. Research is needed to determine ranges and methods for most laboratory animal species.

Tramadol

Tramadol[2] is a centrally acting synthetic analgesic used to treat postoperative and chronic pain in humans. It has a multimodal action: it is an opioid receptor agonist and it inhibits norepinephrine and serotonin reuptake from neurons where those amines are released, including in the spinal cord where both norepinephrine and serotonin can contribute to the modulation of nociception (Grond and Sablotzki 2004). An active (M1) metabolite of tramadol binds with high affinity to mu-opioid receptors; indeed it has more affinity for the opioid receptor than the parent drug. The use of tramadol has recently increased significantly in veterinary medicine. However, in humans and dogs (and possibly other species) with an inherited

[2] Draft FDA guidance on tramadol is available at www.fda.gov/downloads/Drugs/GuidanceComplianceRegulatoryInformation/Guidance/ucm090703.pdf (accessed July 28, 2009).

deficiency of cytochrome P450 2D6 the M1 metabolite is not produced and the drug is therefore less effective (KuKanich and Papich 2004; Stamer et al. 2003). Oral tablets as well as a combination with acetaminophen are currently commercially available in the United States, whereas the parenteral formulation is not. The inability to administer tramadol by injection may limit its usefulness in animals, as clinical experience has shown that its bitter taste makes it aversive to dogs, cats, primates, and rats. The parenteral formulation, if obtained, can be given by intramuscular, intravenous, subcutaneous, or intraperitoneal injection. Affaitati and colleagues (2002) found that subcutaneous injection of tramadol in a rat model of ureteral calculosis reduced signs consistent with visceral pain. Tramadol analgesia is enhanced when combined with other types of analgesics (KuKanich and Papich 2004). Doses in dogs and cats, and possibly in rats and mice, may be estimated from published pharmacokinetic data and dose response studies, but in general more research on the effects of and methods to administer tramadol is needed for laboratory animal species.

NSAIDs

Nonsteroidal anti-inflammatory drugs (NSAIDs) are used to treat postoperative chronic and inflammatory pain in humans and animal species. NSAIDs are classified as "antihyperalgesics" rather than as true analgesics since they do not increase the pain threshold in normal, uninjured subjects (Ghilardi et al. 2004; Yaksh et al. 1998). This very useful class of drug inhibits various isoforms of cyclooxygenase (COX), thus reducing the production of prostaglandins (Samad et al. 2002), a key component of the inflammatory reaction. Prostaglandin inhibition either at the site of tissue injury or centrally at the spinal cord can modulate pain. At least three isoforms of COX have been identified and drugs that selectively inhibit the various isoforms have been created in the search for an effective drug with few side effects. Commercially available selective COX-2-inhibiting NSAIDs are very important drugs for pain management in dogs.

Despite increased cardiovascular risk in adult human populations, adverse cardiovascular effects of COX-2 selective NSAIDs have not been reported in veterinary species. However, because of their inhibition of COX isoforms, NSAIDs are capable of causing injury through their effects on various organ systems. These effects include gastric ulceration and perforation, acute renal failure, and decreased coagulation due to inhibition of platelet aggregation.

In animals for which therapeutic dose ranges have been determined, NSAIDs can be used as relatively long-acting (12-24 hour) agents for momentary, procedural, and persistent or chronic pain. They can also be combined with other analgesics in a multimodal approach. Pharmacokinetics are known for some NSAIDs in dogs, ruminants, horses, rats, mice, and several species of fowl (Baert and De Backer 2003; Busch et al. 1998; Engelhardt et al. 1996; Lascelles et al. 2007; Lees 2003; Lees et al. 2004; Tranquilli et al. 2007). Effectiveness has been demonstrated

- in soft tissue models of pain in dogs, cats, rats, and mice (Kroin et al. 2006; Lascelles et al. 2007; Leece et al. 2005; Roughan and Flecknell 2001, 2004; Whiteside et al. 2004; Wright-Williams et al. 2007);
- for orthopedic pain in dogs, cats, fowl, mice, rats, and horses (Barton et al. 2007; Danbury et al. 2000; Hocking et al. 2005; El Mouedden and Meert 2007; Lascelles et al. 2007; Luger et al. 2002; Valverde and Gunkel 2005);
- in rat neuropathic pain models (Lynch et al. 2004); and
- in visceral pain models in mice and rats (Engelhardt et al. 1995; Millecamps et al. 2004; Miranda et al. 2006).

The efficacy of NSAIDs in nonmammalian, nonavian species is unknown.

Local Anesthetics

Local anesthetics are effective both in awake or sedated animals to reduce momentary, non-tissue-damaging pain (e.g., needle biopsy) and in anesthetized animals as supplements during surgical procedures (Robertson 2005; Valverde and Gunkel 2005; White 2005). Their effect is due to the reversible binding of neuronal sodium channels and the ensuing inhibition of neural conduction (Valverde and Gunkel 2005); by decreasing sensory input, local anesthetics inhibit peripheral and central sensitization (White 2005).

The chief disadvantages of local anesthesia/analgesia are that certain techniques (e.g., epidural or regional nerve blocks) require technical expertise and even long-acting local anesthetics have relatively short durations of effect (4-6 hours, depending on the site). Local anesthetics also have antimicrobial and anti-inflammatory properties, which may limit the benefits of their intermediate-term use in studies of inflammation (Cassuto et al. 2006). Potential advantages of local anesthetic use include the opportunity to reduce general anesthetic doses (thus reducing anesthetic-induced cardiovascular depression), comfortable awakening from surgery, and excellent postoperative analgesia without unwanted side effects (e.g., sedation and ileus; Robertson 2005; Valverde and Gunkel 2005; White 2005).

Local anesthetic techniques have been reported for most domestic animals and, if not, may be extrapolated from studies done in rodents.

NMDA Receptor Antagonists

Ketamine, a dissociative anesthetic, and several other unrelated drugs (such as memantine) modify nociceptive signal transmission and block the induction and maintenance of central sensitization by blocking N-methyl-D-aspartate (NMDA) receptors (Himmelseher and Durieux 2005). As a "central sensitization modulator," ketamine acts by reversing allodynia, hyperalgesia, and opioid tolerance rather than as an analgesic. However, at low and subanesthetic doses it exhibits analgesic properties, hence its use for the management of pain in a variety of situations (Visser and Schug 2006). For example, studies in animals (horses, dogs, mice, rats) and humans have shown that low doses of ketamine (and other NMDA receptor antagonists) reduce the required concentration of inhalant anesthetics during surgical procedures, contribute to opioid sparing, prevent opioid tolerance, reduce acute somatic and visceral pain, and aid in the treatment of neuropathic pain (Anand et al. 2007; De Kock and Lavand'homme 2007; Himmelseher and Durieux 2005; Knotkova and Pappagallo 2007; Lu et al. 2003; Muir et al. 2003; Price et al. 1996; Richebe et al. 2005; Strigo et al. 2005; Valverde and Gunkel 2005).

Ketamine is extensively used in anesthetic regimens for animals. High doses in combination with another anesthetic drug (e.g., xylazine) are commonly used to anesthetize a variety of laboratory animals (particularly rodents). The optimum duration of ketamine administration for effective postsurgical pain management is unknown, although it may be that intraoperative dosing with extension into the postanesthetic period is optimal (Himmelseher and Durieux 2005). Further studies are needed to determine whether this speculative benefit of ketamine is valid.

α_2-Adrenoreceptor Agonists

Drugs that act on α_2-adrenergic receptors (α_2-adrenoreceptor agonists) in the dorsal horn of the spinal cord produce analgesia accompanied by cardiovascular depression and sedation (Kamibayashi and Maze 2000). In human patients those side effects can be particularly limiting, but in stable veterinary patients some degree of sedation is often useful. In equine and small animal practice "microdose" administration of α_2-adrenoreceptor agonists (detomidine and medetomidine in horses, medetomidine in dogs and cats) is used clinically to enhance pain relief as well as reduce anxiety in trauma and surgical patients. One of the major advantages of this class of drugs is the ease with which the sedative effects are reversed (although the reversal also applies to any analgesic effect). Inhibition of both postoperative/postprocedural and neuropathic pain by α_2-adrenoreceptor agonists has been shown in many animal models, but the clinical consequences of this outcome are unknown (Murrell and Hellebrekers 2005). High doses of α_2-adrenoreceptor agonists used as presurgical sedatives or in anesthetic regimes may confer a degree of perioperative analgesia in laboratory animals, but this remains to be demonstrated.

Unconventional Analgesics: Antiepileptic Drugs

Anticonvulsants or antiepileptic drugs (AEDs) act to reduce neuronal hyperexcitability, and there is currently intense interest in their use for both surgical "protective premedication" and chronic cancer pain states as they appear to have antihyperalgesic properties. Two AEDs, gabapentin and a new chemically related congener, pregabalin, are approved for chronic pain management in humans (particularly neuropathic pain such as postherpetic neuralgia), and gabapentin is being investigated for treatment of surgical pain as well (Dahl et al. 2004; Mathiesen et al. 2007). There are numerous reports of the efficacy of these two AEDs to reduce the sensitized state of postoperative/postprocedural and persistent pain in animal models (e.g., Blackburn-Munro and Erichsen 2005). Gabapentin is synergistic with other analgesics and is currently used (empirically) for chronic pain management in dogs and cats.

The Role of Anxiolytic Drugs in Pain Management

Fear or anxiety-related stress may enhance pain (see Chapter 1 and "Modulatory Influences on Pain: Anxiety, Fear, and Stress" in Chapter 2). Studies have shown that pain is both a cause of and worsened by anxiety (Linton 2000; Morley et al. 1999; Munro et al. 2007; Panksepp 1980; Perkins and Kehlet 2000; Ploghaus et al. 2001). Drugs with anxiolytic properties in animals include phenothiazines, which can be either short- (e.g., acepromazine) or long-acting (e.g., zuclopenthixol, fluphenazine); butyrophenones (e.g., azaperone, haloperidol); and benzodiazepines (e.g., diazepam). Evidence is mounting that antiepileptic drugs may also have anxiolytic properties at lower doses than those that provide analgesia or antiseizure effects (Munro et al. 2007). Measures to reduce fear and anxiety, whether pharmacological or nonpharmacological, should be considered important in the reduction of pain.

Confounding and Beneficial Effects of Anesthetics and Analgesics

The laboratory animal, whether used as a whole animal or as a source of tissue for in vitro preparations, is susceptible to an array of influences on its normal function. Clearly, any drug-induced or unintended physiologic state (e.g., pain, dehydration, acid base imbalance) in an animal model may affect the outcome. Anesthesia and analgesia are integrally involved in the humane care of laboratory animals, but they are also essential tools that can contribute to the success of an experiment. It is essential that the investigator understand how such drugs may affect an animal so that experiments can be designed to minimize, balance, or control for confounding variables. Selected situations are illustrated below.

Neurotoxicity

The developing CNS is exquisitely sensitive to its internal milieu (Bhutta and Anand 2002). Although the immature brain undergoes some degree of baseline neurodegeneration by apoptotic processes as part of normal development (Kuan et al. 2000), exposure to certain drugs (including those for therapeutic or anesthetic use) or stressors (e.g., noxious stimuli, maternal deprivation, hypoglycemia, hypoxia, ischemia) during this critical window leads to pathological neurodegeneration.

The neurotoxic effect of CNS depressants on the developing brain was heralded by the Olney group, which reported accelerated neurodegeneration in rat pups exposed to NMDA receptor antagonists, γ-aminobutyric acid (GABA) agonists, and anticonvulsant drugs (Bittigau et al. 2002; Ikonomidou et al. 1999, 2000). Similarly, Slikker and colleagues (2007) reported increased neurodegeneration in fetal and postnatal (day 5) rhesus monkeys exposed to ketamine for 24 hours, but not after 6 hours, confirming that ketamine dose and duration both play important roles in ketamine-induced neurodegeneration (Hayashi et al. 2002; Anand et al. 2007). Since many anesthetic drugs or adjuncts are either NMDA receptor antagonists (e.g., ketamine) or $GABA_A$ receptor agonists (benzodiazepines, barbiturates, and volatile anesthetics), prolonged administration of these drugs during the perinatal period may have significant consequences on brain development and function (Loepke and Soriano 2008).

Initially, it was thought that anesthetic and anticonvulsant drugs (and ethanol) simply accelerate the normal "pruning" or apoptotic process. However, permanent changes in brain histology and in behavioral and locomotor performance have recently been reported in mature rats exposed to isoflurane and midazolam during infancy (Jevtovic-Todorovic et al. 2003). The revelation that anesthetic drugs are neurotoxic suggests a link to the neurodegenerative sequelae of fetal alcohol syndrome. However, no phenotype of "fetal or neonatal anesthesia syndrome" has been demonstrated (Soriano et al. 2005). This issue is of paramount interest to pediatric anesthesiology and intensive care researchers who study the safety of fetal and neonatal anesthesia (Anand and Soriano 2004; Todd 2004).

Anesthetic-induced neurotoxicity is not limited to the young (Anand 2007); several investigators have demonstrated both transient and long-term cognitive dysfunction in older rats (12-24 months). Exposure to isoflurane and nitrous oxide resulted in improved spatial memory in young rats but impaired it in aged rats for at least 3 weeks, indicating that anesthetics can influence memory for much longer than the pharmacokinetic properties of the drug suggest and may adversely affect memory processes in the elderly (Culley et al. 2003, 2004). Furthermore, isoflurane induces beta-amyloid protein deposition and apoptotic cell death, similar to the neurodegenerative process observed in Alzheimer's disease (Xie et al. 2007). Accordingly, research that requires administering anesthesia prior to assessment of cognitive function should take into account the long-term effects of anesthetic exposure.

Neuroprotection

Anesthetic drugs have been shown to impart neuroprotective effects as well. In contrast to the neurotoxic effect of NMDA receptor antagonists described above, ketamine and memantine also protect neurons from excitotoxic injury. Anand and colleagues (2007) examined the effect of a low (sedative) dose of ketamine on P7 rat pups subjected to repetitive inflammatory pain (such pain increases neuronal excitation and cell death in developmentally regulated cortical and subcortical areas). Ketamine at a dose of 5 mg/kg (i.e., a quarter of the dose that induces neurodegeneration in unstimulated rat pups) attenuated cell death and provided some degree of neuroprotection. Memantine has been shown to reduce cognitive decline in patients with Alzheimer's disease (Tariot et al. 2004). Xenon, an inert gas that is a weak NMDA receptor antagonist and thus displays some anesthetic properties, and dexmedetomidine, an α_2-adrenoreceptor agonist, decreased the infarct volume in P7 rat pups after experimental focal cerebral ischemia (Ma et al. 2007). Furthermore, the coadministration of xenon prevented isoflurane-induced neurodegeneration during a 6-hour exposure to 0.75% isoflurane in neonatal rats (ibid.).

Reports also indicate that the inhalant anesthetic isoflurane is neuroprotective during hypoxia-ischemia in in vivo and in vitro animal models of the developing brain (Loepke et al. 2002; McAuliffe et al. 2007; Zhao and Zuo 2004) and during focal cerebral ischemia. Sakai and colleagues (2007) demonstrated that isoflurane provided long-term protection (for 1 month) in terms of reduced injury after experimental stroke (for a review of the preconditioning neuroprotective effects of inhalant anesthetics see Wang et al. 2008).

Cardioprotection

During cardiac surgery, or in models designed to study cardiovascular disorders, consideration is often given to the fact that ischemia and reperfusion of the ischemic heart can induce myocardial injury and cell death.

Anesthetic or analgesic drugs used during procedures may exert important effects on the models (Riess et al. 2004; Suleiman et al. 2008). Brief episodes of nonlethal ischemia (which may be intentional or unintentional, as in the case of excessive depth of anesthesia, hypotension, or tachycardia) activate mechanisms that lead to protection of cardiac myocytes from further injury (Post and Heusch 2002; Suleiman et al. 2008; Weber et al. 2005). This phenomenon is known as cardiac preconditioning. Protection against myocardial damage after ischemic insult is also a well-known effect of volatile (inhalant) anesthetics; and opioids (those acting at delta-opioid receptors) and potentially other anesthetics or analgesics may have similar properties (Barry and Zuo 2005; Peart et al. 2005).

The timing of drug administration (before, during, or after ischemia) determines whether an intervention has an effect and if so the degree of cardioprotection (Schipke et al. 2006; Weber et al. 2005). Other anesthetic agents (e.g., propofol, ketamine, thiopental) may also have myocardial preconditioning or protective effects (Suleiman et al. 2008). The molecular events surrounding myocardial damage and conditioning effects of ischemia and drug therapy have been reviewed (Peart et al. 2005). The triggering of a pro-inflammatory state by surgery, anesthetics, or devices, as well as potential anti-inflammatory effects of drugs may all play a role in the outcome of cardiac procedures (Suleiman et al. 2008). It appears that the method with which animals are anesthetized and treated for pain may influence experimental findings in cardiac-surgical or cardiac-disease models, but these interactions are extremely complex and not fully delineated.

Immunosuppression and Reduction of the Inflammatory Response

Experimental in vivo models of cancer, infectious diseases, trauma (including surgery), hypoxia, ischemia, or toxicity activate a complex orchestration of inflammation, cellular defenses, and repair mechanisms. Inflammation is part of the immune response, a "first responder" that protects the animal from invading organisms or insults and modulates cellular and homeostatic events. Most, if not all, modern anesthetic agents can alter certain inflammatory markers of immune function in in vitro and in vivo models both in humans and in animal models (Galley et al. 2000; Homburger and Meier 2006; Kona-Boun et al. 2005; Lemaire and van der Poll 2007; Schneemilch et al. 2005). However, general anesthesia is not the primary determinant of immune status, for its effect is substantially augmented by the concomitant stress response and surgical tissue injury. Immune function is influenced by an interaction between doses and timing as well as nondrug factors such as pain, psychologic state, perioperative blood loss, or hypothermia (Galley et al. 2000; Homburger and Meier 2006; Padgett and Glaser 2003; Vallejo et al. 2003). Indeed, some authors suggest that the actual clinical significance of anesthetic-induced immunosuppression is minor (Galley et al. 2000).

Analgesic agents also affect immune function. But although opioids cause immunosuppression this effect may be highly dependent on the situation. Morphine, for example, induces changes in natural killer cell activity, inflammatory cytokine production, and mitogen-induced lymphocyte proliferation that lead to immunosuppression in both in vitro and in vivo models (Page 2005; Roy et al. 2005). Conversely, in the context of surgical or cancer pain models, treatment with various opioid analgesics (fentanyl, morphine, tramadol) paradoxically seems to improve immune system function, by inhibiting metastatic spread of cancer cells and limiting tumor growth (Gaspani et al. 2002; Page et al. 2001; Sacerdote et al. 2000; Sasamura et al. 2002). In animal models, there appear to be differences between opioid agents, with buprenorphine contributing less to immune dysfunction than fentanyl (Franchi et al. 2007; Martucci et al. 2004). There is speculation that opioids may be less immunosuppressive when they are given in the context of pain than in in vitro or in vivo animal models without pain (Page 2005). The immune system effects of perioperative and/or chronic opioids may therefore depend on the specific opioid (e.g., buprenorphine versus fentanyl) and on the relationship between the dose and the amount of pain.

Other analgesics (e.g., local anesthetics, ketamine) have been shown to play a role in modulating inflammatory or immune system function (Beilin et al. 2007; Cassuto et al. 2006; Homburger and Meier 2006). Experiments that focus on inflammation or immune function as an outcome measure should require in-depth knowledge of the relevant contributions of analgesic and anesthetic drugs. Implication or exclusion of analgesic drugs in the experimental design may be appropriate *only* if other factors that affect the inflammatory response or immune function are well controlled.

Nonpharmacologic Management of Pain

Most nonpharmacologic methods to treat pain predominantly address acute and chronic musculoskeletal pain conditions. Techniques may include electrostimulation, local tissue cooling (cryotherapy), heat, and manual therapy. These techniques are time- and thus cost-intensive and may require specialized training. As research into mechanisms and efficacy continues, the reader is encouraged to search for more up-to-date information.

Electrotherapy and electrostimulation techniques are commonly used to treat pain in humans; modalities include transcutaneous electrical nerve stimulation (TENS), interferential therapy, and electroacupuncture. Although animal models have shown a reduction in primary and secondary hyperalgesia after TENS treatments (Ainsworth et al. 2006; Hingne and Sluka 2007), definitive support for the use of TENS in laboratory animal medicine is lacking. Similarly, although acupuncture is commonly used in the management of human musculoskeletal pain (e.g., osteoarthritis, intervertebral disk disease, rheumatoid arthritis), its effectiveness in managing animal pain has not been adequately studied.

Cryotherapy, typically used in situations of brief injury or active inflammation, is probably one of the most easily applicable techniques in a laboratory animal facility, requiring little training or cost. It can be achieved using crushed ice, frozen gel packs, frozen alcohol/water slushes, specialized cryotherapy units, or cold sprays. However, despite its common use with acute injury, there is little definitive evidence of a pain-relieving benefit for acute or chronic pain (Greenstein 2007). It is important to seek further guidance prior to its use.

The benefit of therapeutic heating, produced by either deep (therapeutic ultrasound) or superficial (moist hot packs, immersion baths, infrared light) methods has not yet been proven for laboratory animals.

Manual modalities include joint manipulation/mobilization and massage. Although studies have demonstrated the efficacy of such therapies in humans, physical manipulation (e.g., chiropractic adjustment) in animals requires advanced training and is still poorly understood. Limited forms of massage therapy for animals may enhance comfort in joint or muscle pain (especially in animals immobilized by their physiologic state or in restraint devices) although basic training is necessary. At the very least, massage or other hands-on therapies promote bonding between handler and animal in amenable species, can be calming, and may further accustom animals to being touched.

Other Nonpharmacologic Measures to Improve Comfort

Environmental and physical factors can exacerbate pain that results from disease or injury. Nonpain sources of discomfort or distress, such as nausea, hunger, dehydration, dizziness, or weakness, should always be considered (McMillan 2003). Changes in environment, such as deeper or softer bedding, alternative feeding strategies, dim lights, or warmer temperatures may improve the comfort of debilitated animals or those with lower pain thresholds. The use of other appropriate supportive measures, such as parenteral fluid supplementation and wound care, are critical adjuncts to optimize animal comfort and welfare.

PRACTICAL APPLICATIONS AND CONSIDERATIONS FOR PAIN MANAGEMENT

Minimization of Momentary, Non-Tissue-Damaging Pain

Procedures of short duration that do not cause significant tissue damage may nonetheless cause transient pain that is aversive to animals. Examples of such procedures include the placement of intravenous catheters; injection or sampling with large gauge needles; removal of staples, sutures, chest tubes, and abdominal drains; and oral gavage. Especially when vigorous movement in response to a painful stimulus is likely, techniques that reduce pain might also reduce the potential for injury from struggling and enhance the accuracy of the procedure as well as the safety of the handler. Pharmacologic measures for the minimization of such brief types of pain include general anesthesia, sedation, and local anesthesia.

In small animals (e.g., rodents, piglets, cats) brief episodes of inhalant anesthesia may be induced via mask or chamber using isoflurane or sevoflurane followed by mask maintenance and recovery in a protected environment. Major advantages of this approach include rapid onset and recovery times as well as multiple administrations without lingering drug effects. However, in certain species (e.g., ruminants) mask induction of inhalant anesthesia is not appropriate due to risk of regurgitation and aspiration or, in very large animals and primates, size and restraint challenges.

Sedation and neuroleptanalgesia may be useful for minimizing minor procedural pain; examples of such uses include the administration of intravenous propofol during aspiration of vitreous fluid from the eyes of dogs, or a combination of opioid/α_2-adrenoreceptor agonist administration before ultrasound-guided needle biopsy in dogs or ruminants. Animals under prolonged sedation may require additional support (e.g., observation, thermal supplementation, protection from physical harm during recovery) but with many current techniques recovery may be hastened with pharmacologic reversal of the drug (e.g., opioid receptor or α_2-adrenoreceptor antagonists). However, pain and distress may return when pharmacologic reversal of these drugs is used to awaken animals, so in some cases it may be preferable to allow spontaneous recovery.

Topical application of local anesthetic preparations or a short, 1- to 2-minute application of ice or vapocoolant spray to the site may greatly reduce the pain of injection or other superficial pain-producing procedures; studies in cats and humans show reduced pain during minor procedures following topical application of local anesthetics (Gibbon et al. 2003; Howard 2005; Luhmann et al. 2004; Wagner et al. 2006; Weise and Nahata 2005). And local application of lidocaine, for example, may even reduce pain produced by more invasive techniques, such as biopsy or bone marrow aspiration. Disadvantages of topical local anesthetics include prolonged (20-60 minutes) delay in effect when applied to intact epithelium, propensity for removal by the animal, lack of information about concentration (dose) and efficacy for many animal species, and expense. Furthermore, the injection of local anesthetics often causes an acute burning sensation (30 sec to 1 min), which can be alleviated by buffering the drug solution with sodium bicarbonate (Burgher and McGuirk 1998; Burns et al. 2006), administering sedation or a topical anesthetic, or applying local cooling with ice or vapocoolants (Luhmann et al. 2004; Ong et al. 2000). Topical lidocaine gel or solution warmed to body temperature is well absorbed through mucosal surfaces (but not through intact mammalian skin) and is an effective means to reduce the pain of urethral or nasal cannulation and of many ocular procedures.

The benefit of nonpharmacologic measures to minimize the brief but potentially distressing pain of minor procedures in animals is frequently underestimated, but there is evidence of its effectiveness in human neonatal and adult medicine (Golianu et al 2007; Houck and Sethna 2005). Techniques such as topical cooling, physical distraction, and training might be easily incorporated to reduce brief aversive pain.

Ice

Studies in human pediatric and adult medicine show that the application of ice reduces brief pain associated with intramuscular and intradermal injection of drugs and local anesthetics (Farion et al. 2008; Hasanpour et al. 2006; Hayward et al. 2006; Kuwahara and Skinner 2001; Yoon et al. 2008). Although the usefulness of ice application has not been studied in laboratory animals, it should be considered a helpful method for the alleviation of brief pain in laboratory animals.

Physical Distraction and Training

Examples of physical distraction techniques to manage brief pain in animals include the use of a twitch, snare, or "shoulder roll" in horses and livestock, the gentle scruffing of cats, and the gentle pinch of a skin fold in dogs, all of which presumably activate mechanoreceptors and modulate nociceptive transmission. The mechanism of action is not yet clear, but suggestions include the release of endogenous opioid peptides in response to stress or the "gate control theory" (for more information see Lagerweij et al. 1984 or Dickenson 2002).

Randomized controlled studies in humans support the application of various mechanical stimuli for reducing procedural pain. Methods examined include pressure to the site of intramuscular injection of a large volume (Barnhill et al. 1996), and leg massage and facilitated tucking or swaddling prior to heel stick in preterm infants/neonates (Corff et al. 1995; Howard 2005; Jain et al. 2006). However, the amount of pressure applied can make the difference between a pain-reducing and a pain-producing stimulus; a useful guideline for large laboratory animals is that pressure not exceed that which the handler could apply comfortably to his or her own body.

Positive reinforcement training of certain socialized species can greatly reduce the need for forcible restraint during brief painful procedures. Animals acclimated to injection or venipuncture or trained to enter a restraint device (e.g., a chair or sling; Laule et al. 2003; Rennie and Buchanan-Smith 2006; Wolfensohn 2004) may willingly submit to mildly painful procedures in return for a reward (e.g., food, release, physical contact). In contrast, the stress of restraint and/or separation from cage or herd mates may increase fear and anxiety, which in turn can enhance pain (see Chapter 2). For animals acclimated to handling, the presence of a familiar individual (human or conspecific) is often beneficial, and soft verbal encouragement from relaxed, nonthreatening handlers is arguably an important stress reduction measure.

Interventions for Postoperative/Postprocedural and Chronic Pain

This section deals primarily with the management of pain generated by procedures that cause tissue damage (e.g., surgery) and by disease-related and chronic conditions. Considerations regarding pain-related research are discussed in Appendix A, while Box 1-1 defines the various categories of pain as used in this report.

Postprocedural and Postsurgical Pain

Substantial tissue damage from surgery or other procedures causes postprocedural pain that increases as inflammation develops in the injured tissues. The intensity of postoperative/-procedural pain usually peaks within 4 to 24 hours, after which, as tissues heal, it subsides and resolves at a variable rate dependent on several factors but principally on the extent of tissue insult. The mainstay of management of postoperative/-procedural pain of moderate to severe intensity in both human and veterinary clinical medicine is systemic administration of opioid receptor agonists (e.g., morphine, fentanyl) or mixed agonists/antagonists (e.g., buprenorphine). As previously discussed, NSAIDs can be effective for the management of mild to moderate pain; however, because they lack true analgesic efficacy, they are frequently combined with opioids and other drugs. Other analgesic or adjunct drugs (see below) commonly used to manage postoperative/-procedural pain include local anesthetics, ketamine, α_2-adrenoreceptor agonists, and, increasingly, tramadol and gabapentin. Cryotherapy is an example of a potentially beneficial nonpharmacologic adjunct to analgesia.

One standard postoperative care approach following very painful surgeries is the initial administration of a high-efficacy opioid receptor agonist to provide surgical-level analgesia for a period of time, followed by a mixed opioid agonist/antagonist (e.g., buprenorphine) or other drug (e.g., tramadol). Alternatively, adoption of a multimodal analgesic regime may be appropriate (e.g., opioid-NSAID, opioid-ketamine, or some other combination). As the intensity of pain decreases, the pain management strategy (e.g., type and frequency of analgesic drug administration) can be modified. Following a change in analgesic strategy, observations for effectiveness must continue. The last step is to taper these high-efficacy opioid follow-on strategies to a single agent as the intensity of pain lessens. Similarly, when analgesics are discontinued altogether, observations must continue regularly, albeit less frequently, to determine whether termination of pain management is appropriate. The time course of postoperative/-procedural pain may vary considerably not only between species but also between individuals.

There is considerable concern that improperly managed postoperative/-procedural pain can evolve into much longer-lasting, even chronic pain. Drugs that function primarily as antihyperalgesics (e.g., NMDA receptor antagonists and COX-2 inhibitors) are under evaluation to prevent what is sometimes referred to as the chronification of pain (Samad et al. 2001, 2002).

Sickness Syndrome

An unintended and underappreciated consequence of invasive procedures is "sickness syndrome." This syndrome occurs when animals are exposed to potent stimulators of the immune/inflammatory response (e.g., endotoxins, antigenic vaccines, certain cancer states, CNS trauma, reperfusion injury, clinical sepsis), as the resulting proinflammatory cytokines can "facilitate" or enhance pain (Cleeland et al. 2003; Romanovsky 2004; Watkins and Maier 2005; Wieseler-Frank et al. 2005). In addition to fever, the animal exhibits generalized clinical signs of hyperalgesia, malaise, inappetence, somnolence, and other signs that may have evolved as a protective mechanism to induce the animal to rest or sleep (Dantzer and Kelley 2007; Wieseler-Frank et al. 2005). In "sickness syndrome" cytokines can also activate glia in the CNS and contribute to the maintenance of generalized central sensitization (Wieseler-Frank et al. 2005). Because many laboratory animal models include some degree of strong immune stimulation associated with the above conditions, it is important to appreciate that sick animals may be more sensitive to external noxious and nonnoxious stimuli. Interventions to reduce these hyperalgesic states are experimental and include strategies to reverse glial activation (see Shäfers et al. 2004;

Watkins and Maier 2005). A decision to withhold analgesics in an apparently "sick" animal should take into account the potentially significant impact of "sickness syndrome," and in any case nonpharmacologic methods to manage pain—such as a protective environment (shelter, dim light, warmth, bedding), protection from conspecifics, and "hospice" husbandry measures—are strongly recommended.

Preemptive Analgesia

The typical approach to treating postoperative pain, whether in animals or humans, is to give analgesics during or immediately after surgery, but the possibility that treatment before surgery can influence postoperative pain has received considerable attention following Woolf's observation of central hyperexcitability associated with postinjury pain (Woolf 1983). Bach and colleagues (1988) reported that aggressive analgesic treatment (daily morphine administration to the spinal cord) before limb amputation in humans significantly reduced the development of phantom limb pain in the first year after surgery. This observation has been confirmed, largely in animals, suggesting that the use of "preemptive analgesia" before surgery can reduce the magnitude of hypersensitivity and pain that normally occur after surgery (Bromley 2006; Gonzalez et al. 2000; Lascelles et al. 1995, 1997; Reichert et al. 2001).

The effectiveness of preemptive analgesia is presumed to reflect the prevention or attenuation of peripheral and central sensitization, both of which would normally develop during and after a surgical procedure. Tissue and nerve damage activate and sensitize peripheral nociceptors, awaken sleeping nociceptors, and produce central sensitization (an increase in the excitability of central neurons; see Chapter 2). In some respects, the consequence of central sensitization is the biochemical establishment of a "memory" of the injury, in which activation of the NMDA receptor is implicated. The behavioral consequence of central sensitization is that normally innocuous stimuli can induce pain (allodynia) and noxious stimuli evoke greater than normal pain (hyperalgesia). In the short term, the hypersensitivity that results is adaptive, as it compels the animal to protect the injured part of the body. However, because central sensitization is associated with multiple molecular, structural, and neurophysiological changes in CNS neurons and glia, it may also be maladaptive if these changes persist beyond the period of expected postoperative pain (perhaps becoming independent of the original injury) or contribute to the development of a chronic pain state (Romero-Sandoval et al. 2008; Watkins and Maier 2005; Woolf 2007).

Because tissue injury produces central sensitization, it may seem appropriate to use preemptive analgesia in surgical cases, treating the animals before surgery with drugs that prevent nociceptor and central sensitization.

However, most such drugs are experimental and are rarely, if ever, used in the management of pain, as they are not yet approved for clinical use and indeed may even be contraindicated. There are some exceptions; for example, the NMDA receptor antagonist ketamine, local anesthetics, and NSAIDs, especially COX-2 inhibitors, have been demonstrated to have some utility in rodents.

The discussion above suggests that preemptive analgesia should be considered in the course of regular surgical procedures, provided the drugs do not interfere with the experimental protocol. Unfortunately, the initial enthusiasm for preemptive analgesia in humans has decreased because most evidence does not indicate that it offers significantly greater control of postoperative hypersensitivity and pain than other appropriate postoperative strategies for pain management. Accordingly, the likelihood is small that preemptive analgesia significantly contributes to a reduction of the hypersensitivity and pain that occur in the weeks to months after surgery (Grape and Tramèr 2007). It is therefore not at all clear that its use should be recommended or required in the laboratory. However, although preemptive strategies may not help much, they probably will not hurt. To the extent that they do not interfere with the science that justified the surgical procedure they should be considered, but the evidence for their essential contribution in experimental animals remains limited.

It bears reiterating that animals will experience pain after surgery if they have not received an analgesic either before or during the procedure. If analgesics are not given until after surgery, there will be a delay until the drug reaches effective analgesic concentrations in brain tissue. Decisions about the management of pain in the recovering animal should take into account the properties of the anesthetic drug(s) used, the anticipated intensity and type of pain caused by the procedure, and the interaction of administered analgesics with anesthetics. For example, buprenorphine given during ketamine-medetomidine anesthesia in rats resulted in the death of some animals, but ketamine-medetomidine anesthesia alone did not (presumably the buprenorphine suppressed the CNS/respiration even further; Hedenqvist et al. 2000). NSAIDs and local anesthetics do not generally interfere with opioid-induced CNS depression or the action of other anesthetics, but NSAIDs may take 30 or more minutes to be effective whereas local anesthetics act rapidly. A "sparing" effect of many types of pre- or intraoperative analgesics may enable *and require* a reduction in doses of general anesthesia, which can be a desirable goal as many general anesthetics depress cardiac output and respiratory drive. Thus the timing of initiation of analgesia is important not only for managing pain at recovery but also for reducing the likelihood of transition of postoperative/-procedural hyperalgesia to a chronic state (Dahl and Moiniche 2004; Kissin 2005; Pogatzki-Zahn and Zahn 2006).

Chronic Pain

Chronic pain (persistent or chronic pain is discussed in Box 1-1 and Appendix A) in laboratory animals may develop as a consequence of experimental procedures (e.g., device implantation), induced diseases (e.g., cancer, diabetes), or husbandry problems. Animals in chronic pain may experience constant, episodic, or escalating pain accompanied by "breakthrough" episodes of more severe pain. Manipulations that are minimally painful in healthy animals may cause significant pain in those already experiencing pain; thus, for example, handling or husbandry procedures may be painful and should be modified accordingly (e.g., with the use of less invasive sampling techniques, administration of additional analgesia prior to handling).

When determining treatments, assessment methods, and endpoints, the etiology of chronic pain is important. Chronic pain can be inflammatory, visceral, neuropathic, or cancer-related (Bennett et al. 2006). Drug classes commonly used to manage chronic pain include NSAIDs, opioids, tramadol, antiepileptics, antidepressants, and, to a lesser extent, NMDA receptor antagonists and local anesthetics. Nondrug therapies can also be helpful. Depending on the type of pain, animals may need different dosages or types of analgesia. For example, mice with bone cancer need and can tolerate tenfold higher doses of morphine than mice with inflammatory pain (caused by complete Freund's adjuvant or formalin injection) at a similar location of the body (El Mouedden and Meert 2007; Luger et al. 2002). In mice the CNS-depressant effects of morphine (determined through performance in motor coordination assays) are less of an impediment when pain is greater; in contrast, in rats with bone cancer morphine analgesia was accompanied by sedation (Medhurst et al. 2002). Some analgesic drug classes that are effective for inflammatory and neuropathic pain are not effective in bone cancer models (El Mouedden and Meert 2007; Luger et al. 2002; Medhurst et al. 2002; Shaiova 2006).

When pain is expected to increase over time, the frequency of observations and possible interventions should also increase. The potential for tolerance, dependence, and withdrawal must also be considered when designing pain management strategies.

Consequences of Unrelieved Pain

It is likely that unalleviated pain will influence the outcome of a research project in a number of ways. Significant unrelieved pain is a stressor that, if the animal cannot adapt to it, causes distress and negative physiologic consequences, not the least of which is immune dysfunction (Bartolomucci 2007; Blackburn-Munro and Blackburn-Munro 2001; Carr and Goudas

1999; Padgett and Glaser 2003; Ulrich-Lai et al. 2006), especially with respect to experimental metastatic models (Gaspani et al. 2002; Page et al. 2001; Sasamura et al. 2002). Unrelieved pain also has specific effects on animal behavior (Karas et al. 2008), such as reductions in food and water intake or body weight (a surrogate marker of oral intake) demonstrated in a number of animal models, including rats, mice, rabbits, and swine (Flecknell et al. 1999; Harvey-Clark et al. 2000; Karas et al. 2001, 2007; Liles et al. 1998; Malavasi et al. 2006; Shavit et al. 2005). In many instances, the administration of analgesics reduces the magnitude of these changes (Flecknell et al. 1999; Harvey-Clark et al. 2000; Karas et al. 2001; Liles et al. 1998; Malavasi et al. 2006).

Other adverse effects of pain and the morbidity it causes (e.g., ileus, impaired respiratory function and tissue oxygenation) have been reviewed for human patients (Akca et al. 1999; Anand 1993; Bonnet and Marret 2005; Kehlet 2004; Mattei and Rombeau 2006) and it is likely that similar pain-induced morbidity occurs in animals as well. It is therefore reasonable to argue that pain relief is good not only for animal welfare but also for the quality of scientific data.

Animal Welfare Considerations of Research with Persistent Pain Models

Research on pain as a study subject is described in Appendix A. Because of the painful nature of these models and the underlying assumption that analgesics may interfere with the research outcomes, it is important to consider the following questions:

- Is it possible to objectively assess discomfort and/or spontaneous pain or recognize the differences between these and the implications for animal welfare?
- Is the presence of some (or any) spontaneous pain acceptable in order to meet experimental objectives?

The fact that these models have been developed and are in common use indicates that investigators, IACUCs, and veterinarians agree that the presence of ongoing discomfort and/or spontaneous pain can be acceptable if warranted by experimental objectives. However, methods to assess spontaneous pain (or "pain at rest") have not been universally validated. Professional judgment and limited evidence based on the monitoring of a variety of animal behaviors (e.g., food and water ingestion, sleeping, nocturnal activity, sexual activity; see Chapter 3) suggest that, with the exception of the immediate postprocedural period, rodents grow and gain weight and appear to resume normal species-specific behaviors in models of inflammatory, incisional, and even most peripheral neuropathic pain (for references related to specific pain models see Appendix A). But most of these studies have been relatively unsophisticated (e.g., measures are subjective, observers are not blinded), and most species' spontaneous pain-related behaviors have not been studied.

Moreover, the question arises whether to treat what appears to be spontaneous pain in such models, as central nervous system and invasive cancer pain models typically increase in pain intensity and are irreversible. Because drug treatment to reduce pain may interfere with the underlying mechanisms that are the focus of study in these and other models (and thus increase animal use by resulting in an invalid experiment), the provision of humane care without compromising experimental objectives could place the investigator and care providers at odds. Clearly, if the focus of study is the biology of pain, treatment of a presumptive spontaneous condition will interfere with the objectives of the experiment (if, however, reducing a presumptive pain condition does not compromise the goals of the study, then treatment is appropriate). The problem, of course, is that many of the drugs used to treat pain also have effects unrelated to pain, as for example the inhibition of cyclooxygenase with NSAIDs. It is not known to what extent blocking these enzymes by pain-relieving drugs will interfere with the primary objective of the experiment (e.g., tumor development) and there is very little evidence to guide either investigators or care providers.

There is accumulating evidence that many of these models are useful to the study of pain mechanisms and pain management and thus their continued use is valid. However, based on the discussion in Chapter 3 and the approach advocated in US Government Principle #4, animals in persistent pain models should be assumed to be in pain most of the time. They may also experience significant pain upon movement, thus severely affecting their quality of life. Therefore, such studies should be planned conscientiously and judiciously to obtain the maximum amount of data from the minimum number of animals and to use and explore alternatives as much as possible. Further, the committee strongly supports the application of humane endpoints in these studies and refers the reader to Chapter 5 for additional information.

ANALGESIA IN SELECTED NONMAMMALIAN SPECIES

Most information about analgesia is available only for certain mammalian species, and it is reasonable to consider extrapolation between similar mammals. It is beyond the scope of this document to provide the full range of information currently available in veterinary formularies and handbooks of veterinary anesthesia and analgesia. The following sections describe what is known about pain or analgesic treatment in several nonmammalian vertebrates whose use in biomedical research is increasing. Until more information about these species is available, this report can be used as a source of reference for investigators, veterinarians, and animal care personnel.

Fish

It has been reported that the nociceptive sensory system in teleost fish is strikingly similar to the mammalian system and that fish show complicated aversive behavioral and physiological responses to noxious stimuli (Sneddon et al. 2003a,b). These responses were alleviated by morphine (Sneddon 2003), but whether fish are capable of pain perception as opposed to only nociception remains uncertain (Sneddon 2006).

The fish brain is activated during noxious stimulation (Dunlop and Laming 2005; Nordgreen et al. 2007; Reilly et al. 2008a) and there are species-specific differences in response to the same noxious event in common carp (*Cyprinus carpio*), zebrafish (*Danio rerio*), and rainbow trout (*Oncorhynchus mykiss*; Reilly et al. 2008b; also see Chapter 3 and ILAR 2009 for a more extensive discussion of pain perception in fish). Rainbow trout and zebrafish typically respond to noxious events by reducing activity and frequency of swimming; noxious stimulation also causes a rapid rise in respiration rate to almost double that of normal rates. Common carp do not exhibit the same responses, but do show anomalous behaviors associated with loss of equilibrium or "rocking" on their pectoral fins on the substrate. Rainbow trout exhibit these behaviors as well, but in addition rub the stimulated area on the gravel bottom or sides of the tank. Such anomalous behaviors are not observed in zebrafish. Therefore, more species may have to be assessed before reliable criteria can be developed for recognizing pain or discomfort in fish.

There is little information regarding dosage and route of analgesia in fish; only three analgesics—morphine, ketoprofen, and butorphanol—have been assessed so far. Morphine at a wide range of investigational doses (2.5-30 mg/kg i.m.[3]) has been shown to reduce nociceptive responses in rainbow trout at the lowest effective dose of 5 mg/kg i.m. (Sneddon 2003). Butorphanol, investigated in chain dogfish (*Scyliorhinus retifer*) and koi carp (*C. carpio*), was ineffective in the elasmobranch dogfish (dose range 0.25-5 mg/kg; Davis et al. 2006). In contrast, it was effective in diminishing postsurgical changes in behavior and physiology in the teleost bony carp at a dose of 0.4 mg/kg i.m. (Harms et al. 2005). The NSAID ketoprofen (1-4 mg/kg) also had no effect in the dogfish; however, both ketoprofen and butorphanol were given via immersion, and drug uptake through the gills may not have occurred since morphine uptake via this route of administra-

[3]The doses of analgesic drugs discussed in the text are for investigational not clinical use unless otherwise indicated.

tion is quite time consuming (Newby et al. 2006). In contrast, ketoprofen appeared to reduce inflammation of the muscles in koi carp but some aberrant postoperative behaviors were still observed after administration (2 mg/kg i.m.). Much research is needed in this area to determine optimum doses and efficacy of different analgesics.

Amphibians

Basic research has significantly delineated the anatomy, mechanisms, and regulation of pain in the Northern grass frog, *Rana pipiens*, and this species has been proposed as a model for opioid research (Stevens 2004). The analgesic efficacy and duration of action of opioids, α_2-adrenoreceptor agonists, and numerous nonopioid analgesics in amphibians have been reported (Mohan and Stevens 2006; Stevens et al. 2001; Willenbring and Stevens 1997). Species differences in the distribution of nociceptors between *R. pipiens* and rodents have also been described (Stevens 2004). However, there are no reports of clinical studies using objectively established indices of pain in amphibians or of pharmacological studies in either *R. pipiens* or laboratory *Xenopus*. Comparison of limited lethality data in *R. pipiens* suggests that the safety index for these agents is quite narrow (Green 2003). Based primarily on the animal's wiping behavior after the application of acetic acid to its skin, scientists have tested a few analgesic agents and their doses (Terril-Robb et al. 1996; Stevens et al. 2001; Machin 2001; Smith 2007). More studies of specific techniques are needed.

Reptiles

Information about efficacy, pharmacokinetics, and adverse effects of analgesics in reptiles is extremely limited. Indeed, there is no toxicity information to guide local anesthesia/analgesia dosing for reptiles; many authors advise adoption of the dose limits for dogs and cats. In a 2004 survey of 367 veterinary practitioners who treat reptiles, 98% of the respondents indicated that they believed that "reptiles feel pain"; approximately 40% reported the use of empirical or extrapolated methods to prevent or manage pain in their reptiles (Read 2004). The most commonly used drugs were opioids, NSAIDs, and local anesthetics.

Evidence for opioid analgesic efficacy in reptiles is found in less than a handful of reports (Mosley 2005). In a recently published study in the red-eared slider (turtle), Sladky and colleagues (2007) reported that subcutaneous butorphanol administration at 2.8 or 28 mg/kg did not provide analgesia in a thermal latency assay. In contrast, morphine produced long-lasting (24-hour) increases in response latency and concomitant "marked and prolonged" respiratory depression. In green iguanas butorphanol was not shown to reduce isoflurane anesthetic requirements; it also did not adversely affect cardiovascular function (Mosley et al. 2003, 2004). Tuttle and colleagues (2006) investigated the pharmacokinetics of the NSAID ketoprofen in the green iguana and determined that, based on the drug's long elimination half-life, the standard practice of daily dosing might be excessive, although a 10-day course of carprofen or meloxicam administration in the green iguana did not reveal any detrimental effects on the animal's hemogram and chemistry (Trnkova et al. 2007).

Birds

Therapeutic interventions to address pain in birds are based predominantly on studies of fowl (ducks, chickens, turkeys) and various psittacine (parrot) species. The most commonly studied analgesic drugs are NSAIDs and opioids; pharmacokinetic information (to guide dose and duration) is available for certain NSAIDs, but information on the effectiveness of other pain management strategies is limited to professional opinion and best practices. NSAID efficacy studies were conducted with analgesiometric testing (thermal threshold), scoring of clinical parameters (weight bearing, lameness, other behaviors), and/or assessment of self-administration of an analgesic drug. Citations are listed for each type of study described below.

Most clinical parameter and self-assessment testing was conducted in fowl species with naturally occurring or experimental induction of arthritis, or following partial beak amputation (Gustafson et al. 2007). While a dose-response curve is usually a component of these studies, the duration of analgesic action is not easily extrapolated from the results. On the other hand, most of the efficacy studies using analgesiometry were conducted with opioids in perching birds (psittacine species; see text below). These later studies allow an understanding of the duration of the drugs tested, but the type of pain studied (withdrawal threshold to a momentary noxious stimulus) is probably not representative of either postsurgical or chronic pain, so it is important to recognize that the dose and duration information that they convey may differ (i.e., the dose may be higher or lower than needed) in the context of clinical pain in birds. Because class Aves includes species with extremely variable physiological adaptation strategies in which only limited types of pain have been studied, it is probably not feasible to simply extrapolate from the doses described in the literature for the use of analgesics in different bird species. Little to no information on dosing, efficacy, and adverse effects is available for many bird species used in laboratory animal science (e.g., song birds). Studies of surgical analgesia in birds are needed, and beak amputation in fowl may represent an ideal model as it has been reported for birds of varying ages and species.

Studies of opioids in parrots and chickens (Gentle et al. 1999; Paul-

Murphy et al. 1999; Sladky et al. 2007) indicate that opioids acting at the mu-opioid receptor are either not effective in birds or are much less so than in mammals, whereas butorphanol, a kappa agonist opioid with antagonist efficacy at the mu-opioid receptor, is considered the opioid of choice for acute and chronic pain management in birds (Paul-Murphy et al. 1999; Sladky et al. 2007). A chief disadvantage cited for the use of butorphanol in birds is its apparent short duration, requiring frequent redosing. Sladky and colleagues (2007) found long-lasting (up to 5 days) antinociception (to a heat stimulus) and persistent serum concentrations in Hispaniolan parrots given a liposomal encapsulated formulation of butorphanol (10-15 mg/kg) intramuscularly. Morphine has been shown to produce analgesia in certain strains of chickens at much lower doses (15 vs. 100 mg/kg) than in other strains; it is unknown if such strain differences occur in other species (Gentle et al. 1999). A possibly confounding factor is the sedation caused by high doses of some opioids. Opioids are primarily administered intramuscularly; one study showed that intra-articular injection of various doses of morphine, fentanyl, and buprenorphine in chickens did not have an appreciable analgesic effect (ibid.). The best evidence for opioid analgesia in birds currently supports drugs acting at the kappa receptor, but more work is needed to determine optimal administration schedules and specific doses.

NSAIDs are the most extensively studied drugs in birds in terms of pharmacology and efficacy, but their use nonetheless requires piecing together the available information to guide dosing. Carprofen has been shown to reduce clinical signs of both naturally occurring and experimentally induced articular pain in chickens (Danbury et al. 2000; Hocking et al. 2005; Mc Geown et al. 1999); Mc Geown and colleagues (1999) showed that the time lame chickens required to complete an obstacle course was reduced by roughly 50% 90 minutes after intramuscular administration of 1 mg/kg carprofen. In contrast, the minimum effective intramuscular dose of carprofen in a urate model of articular pain in chickens was 30 mg/kg (although mortality was also observed at this dose; Hocking et al. 2005). Naturally lame chickens were found to selectively consume carprofen in feed whereas healthy individuals avoided the medicated feed (Danbury et al. 2000). The authors calculated that the amount of oral carprofen birds consumed to achieve adequate serum levels was approximately 10 times the recommended oral dose for dogs. This study suggests that the dose required depends on the intensity of pain and also points out likely differences in oral bioavailability of this drug in chickens compared with dogs.

A degree of clinical efficacy has been demonstrated for both flunixin and ketoprofen. Hocking and colleagues (2005) found minimum effective doses of intramuscular flunixin (12 mg/kg) and ketoprofen (3 mg/kg) in a urate arthritis chicken model and indicated that the flunixin dose was similar to, and the ketoprofen dose greater than, doses of the drugs recommended for horses and cattle. Although these results establish drug efficacy, they raise concerns about toxicity, including death, and suggest the need to study lower doses.

Machin and colleagues (2001) studied intramuscular ketoprofen (5 mg/kg) in isoflurane-anesthetized mallard ducks and concluded that responses to a noxious stimulus (pressure by clamp) were reduced by ketoprofen 30 to 90 minutes after administration. As NSAIDs have not been reliably found to reduce the minimal alveolar concentration of inhalant anesthetics in mammalian species, the results of this study must be interpreted cautiously. Neither phenylbutazone nor acetaminophen showed analgesic activity in lame chickens (Hocking et al. 2005). Moreover, the efficacy of three anti-inflammatory corticosteroids (betamethasone, dexamethasone, and methylprednisone) was evident from an assay in lame chickens. The authors indicated that the doses used were comparable to those used in mammals for management of pain behaviors (Hocking et al. 2001).

Limited pharmacokinetic data on NSAIDs in birds are available. Baert and DeBacker (2003) compared the pharmacokinetic properties of flunixin, salicylate, and meloxicam in five species of birds (chickens, ostriches, ducks, turkeys, and pigeons) and found that parameters varied by drug and species, and that the typical correlation of elimination half-life with body weight was not evident. Although elimination half-life is important in determining the steady-state serum concentration of a drug, serum concentrations of NSAIDs do not necessarily determine the duration of analgesia. Machin and colleagues (2001) examined plasma thromboxane (TBX) levels following 5 mg/kg of either ketoprofen or flunixin administration in mallard ducks and found that TBX concentrations were suppressed for about 12 hours by both drugs; because intramuscular injection sites in flunixin-treated ducks showed histopathologic evidence of necrosis, the authors caution against this route of injection in these animals.

Adverse effects of NSAIDs in birds have been reported, so these drugs must be used with appropriate caution. Urate accumulation (visceral gout), renal necrosis, and liver damage occur in vultures with oral exposure to the NSAID diclofenac, and the syndrome has been reproduced in domestic chickens (Naidoo et al. 2007). A single dose of ketoprofen (2-5 mg/kg) given to male eider ducks was implicated in deaths from renal damage (Mulcahy et al. 2003). Gastrointestinal damage may also occur in addition to nephropathy, as in mammals, although this is not as well studied.

LIMITATIONS OF AVAILABLE INFORMATION

Pain in animals may not be effectively managed in many situations because of a lack of information about how to recognize and treat it, although controlled studies of analgesia are available for popular veterinary species, primarily for postsurgical and chronic osteoarthritis pain. Some evidence has accumulated in support of pain management strategies for limited types of postoperative and postprocedural pain in some strains of laboratory rodents (e.g., for laparotomy; Karas et al. 2001; Krugner-Higby et al. 2003; Roughan and Flecknell 2001, 2004; Wright-Williams et al. 2007). However, appropriate analgesic treatment for the myriad common surgical approaches in rats, mice, and most laboratory mammals, as well as for chronic, disease-, or cancer-related pain, is mostly extrapolated from research in which pain is the subject of study. In practical terms, analgesia in rabbits, guinea pigs, rodents other than rats and mice, primates, sheep, calves, goats, and swine remains a purely empirical exercise based on anecdote, experience, and best practice. Even less is known about non-mammalian vertebrates, although recent evidence suggests that pain in amphibians and fish may in many ways be analogous to that of mammals (Sneddon 2004; Stevens et al. 1994; Stoskopf 1994). However, this interpretation is controversial because amphibians lack the cerebral and limbic cortical components widely believed necessary for the appreciation of pain (Stevens 2004).

In addition to the absence of scientific evidence, making it difficult to measure the intensity and expected frequency of pain and the efficacy of analgesics in many laboratory species, there are circumstances in which the withholding of analgesic drugs is necessary. One example is pain-related research in which the use of anesthetic and analgesic drugs may not be appropriate because they may interfere with behavioral or other endpoints to be assessed and validated as the focus of the study. Also, many anesthetic and analgesic drugs have inherent properties (protective or toxic) that must be understood and accounted for (e.g., by means of appropriate control groups). Last, anesthetics and analgesics can lead to end-organ injury, either directly through toxic effects or indirectly through impaired vital organ function. Thus, while the specific choice of anesthetics or analgesics is important, so is the manner in which they are used. Investigators should bear in mind that withholding analgesics after surgery or other invasive procedures associated with anticipated moderate to severe pain may confound the results with unwanted variables of immobility, weight loss, and other consequences of stress and pain.

The effective reduction and management of pain in laboratory animals to optimize both their well-being and the quality of the research is still fraught with limitations. However, extrapolation of techniques from other species, accounting for differences in physiology between them, and attention to the vast scientific literature that uses animal models can improve the ability to manage pain in animals in the laboratory.

CONCLUSIONS AND RECOMMENDATIONS

1. Treatment of postprocedural, persistent, and chronic pain requires a basic understanding of its etiology, strategies, and time course. Anticipation of the potential intensity of pain is important in designing the appropriate approach to its prevention or management.
2. The amount of pain experienced by laboratory animals can be reduced through the use of preventive or therapeutic strategies or their combination. Such therapeutic measures include the use of general and local anesthetics, analgesics, and anxiolytics as well as nonpharmacologic methods.
3. Although regulations require treatment for only nonbrief pain, animals subjected to multiple episodes of momentary pain may benefit from measures to alleviate such pain.
4. Limitations to effective pain management include (1) a lack of knowledge of drug effects and doses in many mammalian and, especially, nonmammalian species; and (2) potential confounding effects of analgesics and anesthetics on study variables.
5. In studies where the use of certain analgesics appears to be contraindicated, investigators should be mindful that unwanted variables from pain-induced perturbation of homeostatic mechanisms can affect the animal model.

REFERENCES

ACVA (American College of Veterinary Anesthesiologists). 1998. American College of Veterinary Anesthesiologists' Position Paper on the Treatment of Pain in Animals. Available at www.acva.org/professional/Position/pain.html. Accessed January 5, 2009.

Affaitati G, Giamberardino MA, Lerza R, Lapenna D, De Laurentis S, Vecchiet L. 2002. Effects of tramadol on behavioural indicators of colic pain in a rat model of ureteral calculosis. Fundam Clin Pharmacol 16(1):23-30.

Ainsworth L, Budelier K, Clinesmith M, Fiedler A, Landstrom R, Leeper BJ, Moeller L, Mutch S, O'Dell K, Ross J, Radhakrishnan R, Sluka KA. 2006. Transcutaneous electrical nerve stimulation (TENS) reduces chronic hyperalgesia induced by muscle inflammation. Pain 120(1-2):182-187.

Akca O, Melischek M, Scheck T, Hellwagner K, Arkilic CF, Kurz A, Kapral S, Heinz T, Lackner FX, Sessler DI. 1999. Postoperative pain and subcutaneous oxygen tension. Lancet 354(9172):41-42.

American Heritage Medical Dictionary. 2007. Boston: Houghton Mifflin.

Anand KJS. 1993. Relationships between stress responses and clinical outcome in newborns, infants, and children. Crit Care Med 21:S358-S359.

Anand KJS. 2007. Anesthetic neurotoxicity in newborns: Should we change clinical practice? Anesthesiology 107:2-4.

Anand KJS, Soriano SG. 2004. Anesthetic agents and the immature brain: Are these toxic or therapeutic? Anesthesiology 101(2):527-530.

Anand KJS, Garg S, Rovnaghi CR, Narsinghani U, Bhutta AT, Hall RW. 2007. Ketamine reduces cell death following inflammatory pain in newborn rat brain. Pediatr Res 62(3): 283-290.

Antognini JF, Barter L, Carstens E. 2005. Overview movement as an index of anesthetic depth in humans and experimental animals. Comp Med 55(5):413-418.

Appadu B, Vaidya A. 2008. Monitoring techniques: Neuromuscular blockade and depth of anaesthesia. Anaesth Intens Care Med 9(6):247-250.

Aung HH, Mehendale SR, Xie JT, Moss J, Yuan CS. 2004. Methylnaltrexone prevents morphine-induced kaolin intake in the rat. Life Sci 74(22):2685-2691.

Bach S, Noreng MF, Tjéllden NU. 1988. Phantom limb pain in amputees during the first 12 months following limb amputation, after preoperative lumbar epidural blockade. Pain 33(3):297-301.

Baert K, De Backer P. 2003. Comparative pharmacokinetics of three non-steroidal anti-inflammatory drugs in five bird species. Comp Biochem Physiol C Toxicol Pharmacol 134(1): 25-33.

Barnhill BJ, Holbert MD, Jackson NM, Erickson RS. 1996. Using pressure to decrease the pain of intramuscular injections. J Pain Symptom Manage 12(1):52-58.

Barry U, Zuo Z. 2005. Opioids: Old drugs for potential new applications. Curr Pharm Des 11(10):1343-1350.

Bartolomucci A. 2007. Social stress, immune functions and disease in rodents. Front Neuroendocrinol 28(1):28-49.

Barton NJ, Strickland IT, Bond SM, Brash HM, Bate ST, Wilson AW, Chessell IP, Reeve AJ, McQueen DS. 2007. Pressure application measurement (PAM): A novel behavioural technique for measuring hypersensitivity in a rat model of joint pain. J Neurosci Methods 163(1):67-75.

Beilin B, Rusabrov Y, Shapira Y, Roytblat L, Greemberg L, Yardeni IZ, Bessler H. 2007. Low-dose ketamine affects immune responses in humans during the early postoperative period. Br J Anaesth 99(4):522-527.

Bennett MI, Smith BH, Torrance N, Lee AJ. 2006. Can pain be more or less neuropathic? Comparison of symptom assessment tools with ratings of certainty by clinicians. Pain 122(3):289-294.

Bhutta AT, Anand KJ. 2002. Vulnerability of the developing brain: Neuronal mechanisms. Clin Perinatol 29(3):357-372.

Bittigau P, Sifringer M, Genz K, Reith E, Pospischil D, Govindarajalu S, Dzietko M, Pesditschek S, Mai I, Dikranian K, Olney JW, Ikonomidou C. 2002. Antiepileptic drugs and apoptotic neurodegeneration in the developing brain. Proc Natl Acad Sci U S A 99(23):15089-15094.

Blackburn-Munro G, Blackburn-Munro RE. 2001. Chronic pain, chronic stress and depression: Coincidence or consequence? J Neuroendocrinol 13(12):1009-1023.

Blackburn-Munro G, Erichsen HK. 2005. Antiepileptics and the treatment of neuropathic pain: Evidence from animal models. Curr Pharm Des 11(23):2961-2976.

Bonnet F, Marret E. 2005. Influence of anaesthetic and analgesic techniques on outcome after surgery. Br J Anaesth 95(1):52-58.

Bosgraaf CA, Suchy H, Harrison C, Toth LA. 2004. Diagnosis: Pica secondary to buprenorphine administration. Lab Anim 33(3):22-23.

Bromley L. 2006. Pre-emptive analgesia and protective premedication: What is the difference? Biomed Pharmacother 60(7):336-340.

Bruhn J, Myles PS, Sneyd R, Struys MM. 2006. Depth of anaesthesia monitoring: What's available, what's validated and what's next? Br J Anaesth 97(1):85-94.

Burgher SW, McGuirk TD. 1998. Subcutaneous buffered lidocaine for intravenous cannulation: Is there a role in emergency medicine? Acad Emerg Med 5(11):1057-1063.

Burns CA, Ferris G, Feng C, Cooper JZ, Brown MD. 2006. Decreasing the pain of local anesthesia: A prospective, double-blind comparison of buffered, premixed 1% lidocaine with epinephrine versus 1% lidocaine freshly mixed with epinephrine. J Am Acad Dermatol 54:128-131.

Busch U, Schmid J, Heinzel G, Schmaus H, Baierl J, Huber C, Roth W. 1998. Pharmacokinetics of meloxicam in animals and the relevance to humans. Drug Metab Dispos 26(6):576-584.

Buvanendran A, Kroin JS. 2007. Useful adjuvants for postoperative pain management. Best Pract Res Clin Anaesthesiol 21(1):31-49.

Carr DB, Goudas LC. 1999. Acute pain. Lancet 353(9169):2051-2058.

Carroll GL. 2008. Small Animal Anesthesia and Analgesia. Ames, IA: Blackwell Publishing.

Cassuto J, Sinclair R, Bonderovic M. 2006. Anti-inflammatory properties of local anesthetics and their present and potential clinical implications. Acta Anaesthesiol Scand 50(3): 265-282.

Clark JA Jr, Myers PH, Goelz MF, Thigpen JE, Forsythe DB. 1997. Pica behavior associated with buprenorphine administration in the rat. Lab Anim Sci 47(3):300-303.

Cleeland CS, Bennett GJ, Dantzer R, Dougherty PM, Dunn AJ, Meyers CA, Miller AH, Payne R, Reuben JM, Wang XS, Lee BN. 2003. Are the symptoms of cancer and cancer treatment due to a shared biologic mechanism? A cytokine-immunologic model of cancer symptoms. Cancer 97(11):2919-2925.

Corff KE, Seideman R, Venkataraman PS, Lutes L, Yates B. 1995. Facilitated tucking: A nonpharmacologic comfort measure for pain in preterm neonates. J Obstet Gynecol Neonatal Nurs 24(2):143-147.

Corletto F. 2007. Multimodal and balanced analgesia. Vet Res Commun 37(2):153-160.

Culley DJ, Baxter M, Yukhananov R, Crosby G. 2003. The memory effects of general anesthesia persist for weeks in young and aged rats. Anesth Analg 96(4):1004-1009.

Culley DJ, Baxter MG, Yukhananov R, Crosby G. 2004. Long-term impairment of acquisition of a spatial memory task following isoflurane-nitrous oxide anesthesia in rats. Anesthesiology 100(2):309-314.

Dahl JB, Moiniche S. 2004. Pre-emptive analgesia. Br Med Bull 71:13-27.

Dahl JB, Mathiesen O, Moiniche S. 2004. "Protective premedication": An option with gabapentin and related drugs? A review of gabapentin and pregabalin in the treatment of post-operative pain. Acta Anaesthesiol Scand 48(9):1130-1136.

Danbury TC, Weeks CA, Chambers JP, Waterman-Pearson AE, Kestin SC. 2000. Self-selection of the analgesic drug carprofen by lame broiler chickens. Vet Rec 146(11):307-311.

Dantzer R, Kelley KW. 2007. Twenty years of research on cytokine-induced sickness behavior. Brain Behav Immun 21(2):153-160.

Davis MR, Mylniczenko N, Storms T, Raymond F, Dunn JL. 2006. Evaluation of intramuscular ketoprofen and butorphanol as analgesics in chain dogfish (*Scyliorhinus retifer*). Zoo Biol 25(6):491-500.

De Kock MF, Lavand'homme PM. 2007. The clinical role of NMDA receptor antagonists for the treatment of postoperative pain. Best Pract Res Clin Anaesthesiol 21(1):85-98.

Dickenson AH. 2002. Gate control theory of pain stands the test of time. Br J Anaesth 88(6):755-757.

Dunlop R, Laming P. 2005. Mechanoreceptive and nociceptive responses in the central nervous system of goldfish (*Carassius auratus*) and trout (*Oncorhynchus mykiss*). J Pain 6(9):561-568.

El Mouedden ME, Meert TF. 2007. Pharmacological evaluation of opioid and non-opioid analgesics in a murine bone cancer model of pain. Pharmacol Biochem Behav 86(3): 458-467.

Engelhardt G, Homma D, Schlegel K, Utzmann R, Schnitzler C. 1995. Anti-inflammatory, analgesic, antipyretic and related properties of meloxicam, a new non-steroidal anti-inflammatory agent with favourable gastrointestinal tolerance. Inflamm Res 44(10): 423-433.

Engelhardt G, Homma D, Schlegel K, Schnitzler C, Utzmann R. 1996. General pharmacology of meloxicam—Part II: Effects on blood pressure, blood flow, heart rate, ECG, respiratory minute volume and interactions with paracetamol, pirenzepine, chlorthalidone, phenprocoumon and tolbutamide. Gen Pharmacol 27(4):679-688.

Farion KJ, Splinter KL, Newhook K, Gabourry I, Splinter WM. 2008. The effect of vapocoolant spray on intravenous cannulation pain in children: A randomized control trial. CMAJ 179:31-36.

Fields H. 2004. State-dependent opioid control of pain. Nat Rev Neurosci 5(7):565-575.

Flecknell PA. 2009. Laboratory Animal Anaesthesia, 3rd ed. Elsevier: New York.

Flecknell PA, Waterman-Pearson A, eds. 2000. Pain Management in Animals. Harcourt Publishers.

Flecknell PA, Orr HE, Roughan JV, Stewart R. 1999. Comparison of the effects of oral or subcutaneous carprofen or ketoprofen in rats undergoing laparotomy. Vet Rec 144(3):65-67.

Franchi S, Panerai AE, Sacerdote P. 2007. Buprenorphine ameliorates the effect of surgery on hypothalamus-pituitary-adrenal axis, natural killer cell activity and metastatic colonization in rats in comparison with morphine or fentanyl treatment. Brain Behav Immun 21(6):767-774.

Franks NP. 2008. General anesthesia: From molecular targets to neuronal pathways of sleep and arousal. Nat Rev Neurosci 9:370-386.

Gaertner D, Hallman T, Hankenson F, Batchelder M. 2008. Anesthesia and analgesia for laboratory rodents. In: Fish R, Brown M, Danneman P, Karas A, eds. Anesthesia and Analgesia in Laboratory Animals. San Diego: Academic Press. pp. 239-298.

Galley HF, DiMatteo MA, Webster NR. 2000. Immunomodulation by anaesthetic, sedative and analgesic agents: Does it matter? Intens Care Med 26(3):267-274.

Gaspani L, Bianchi M, Limiroli E, Panerai AE, Sacerdote P. 2002. The analgesic drug tramadol prevents the effect of surgery on natural killer cell activity and metastatic colonization in rats. J Neuroimmunol 129(1-2):18-24.

Gaynor JS, Muir W. 2002. Handbook of Veterinary Pain Management. St. Louis, MO: Mosby.

Gentle MJ, Hocking PM, Bernard R, Dunn LN. 1999. Evaluation of intraarticular opioid analgesia for the relief of articular pain in the domestic fowl. Pharmacol Biochem Behav 63(2):339-343.

Ghilardi JR, Svensson C, Rogers SD, Yaksh TL, Mantyh PW. 2004. Constitutive spinal cyclooxygenase-2 participates in the initiation of tissue injury-induced hyperalgesia. J Neurosci 24(11):2727-2732.

Gibbon KJ, Cyborski JM, Guzinski MV, Viviano KR, Trepanier LA. 2003. Evaluation of adverse effects of EMLA (lidocaine/prilocaine) cream for the placement of jugular catheters in healthy cats. J Vet Pharmacol Ther 26(6):439-441.

Golianu B, Krane E, Seybold J, Almgren C, Anand KJ. 2007. Non-pharmacological techniques for pain management in neonates. Semin Perinatol 31:318-322.

Gonzalez MI, Field MJ, Bramwell S, McCleary S, Singh L. 2000. Ovariohysterectomy in the rat: A model of surgical pain for evaluation of pre-emptive analgesia? Pain 88(1):79-88.

Grape S, Tramèr MR. 2007. Do we need preemptive analgesia for the treatment of postoperative pain? Best Pract Res Clin Anaesthesiol 21(1):51-63.

Green SL. 2003. Postoperative analgesics in South African clawed frogs (*Xenopus laevis*) after surgical harvest of oocytes. Comp Med 53(3):244-247.

Greenstein G. 2007. Therapeutic efficacy of cold therapy after intraoral surgical procedures: A literature review. J Periodontol 78(5):790-800.

Grond S, Sablotzki A. 2004. Clinical pharmacology of tramadol. Clin Pharmacokinet 43(13): 879-923.

Guillou N, Tanguy M, Seguin P, Branger B, Campion JP, Malledant Y. 2003. The effects of small-dose ketamine on morphine consumption in surgical intensive care unit patients after major abdominal surgery. Anesth Analg 97(3):843-847.

Gustafson LA, Cheng HW, Garner JP, Pajor EA, Mench JA. 2007. Effects of bill-trimming Muscovy ducks on behavior, body weight gain, and bill morphopathology. Appl Anim Behav Sci 103:59-74

Harms CA, Lewbart GA, Swanson CR, Kishimori JM, Boylan SM. 2005. Behavioral and clinical pathology changes in koi carp (*Cyprinus carpio*) subjected to anesthesia and surgery with and without intra-operative analgesics. Comp Med 55(3):221-226.

Harvey-Clark CJ, Gilespie K, Riggs KW. 2000. Transdermal fentanyl compared with parenteral buprenorphine in post-surgical pain in swine: A case study. Lab Anim 34(4):386-398.

Hasanpour M, Tootoonchi M, Aein F, Yadegarfar G. 2006. The effects of two non-pharmacologic pain management methods for intramuscular injection pain in children. Acute Pain 8(1):7-12.

Hawk CT, Leary SL, Morris TH. 2005. Formulary for laboratory animals. Blackwell Publications.

Hayashi H, Dikkes P, Soriano SG. 2002. Repeated administration of ketamine may lead to neuronal degeneration in the developing rat brain. Paediatr Anaesth 12(9):770-774.

Hayward SC, Landorf KB, Redmond AC. 2006. Ice reduces needle-stick pain associated with a digital nerve block of the hallux. Foot 16(3):145-148.

Hedenqvist P, Roughan JV, Flecknell PA. 2000. Effects of repeated anaesthesia with ketamine/ medetomidine and of pre-anaesthetic administration of buprenorphine in rats. Lab Anim 34(2):207-211.

Hellyer P, Rodan I, Brunt J, Downing R, Hagedorn JE, Robertson SA (AAHA/AAFP Pain Management Guidelines Task Force Members). 2007. AAHA/AAFP pain management guidelines for dogs & cats. J Am Anim Hosp Assoc 43(5):235-248.

Himmelseher S, Durieux ME. 2005. Ketamine for perioperative pain management. Anesthesiology 102(1):211-220.

Hingne PM, Sluka KA. 2007. Differences in waveform characteristics have no effect on the anti-hyperalgesia produced by transcutaneous electrical nerve stimulation (TENS) in rats with joint inflammation. J Pain 8(3):251-255.

Hocking PM, Robertson GW, Gentle MJ. 2001. Effects of anti-inflammatory steroid drugs on pain coping behaviours in a model of articular pain in the domestic fowl. Res Vet Sci 71(3):161-166.

Hocking PM, Robertson GW, Gentle MJ. 2005. Effects of non-steroidal anti-inflammatory drugs on pain-related behaviour in a model of articular pain in the domestic fowl. Res Vet Sci 78(1):69-75.

Homburger JA, Meiler SE. 2006. Anesthesia drugs, immunity, and long-term outcome. Curr Opin Anaesthesiol 19(4):423-428.

Houck CS, Sethna NF. 2005. Transdermal analgesia with local anesthetics in children: Review, update and future directions. Expert Rev Neurother 5(5):625-634.

Howard RF. 2005. Acute pain management in the neonate. Anaesth Inten Care 6(4):122-124.

Ikonomidou C, Bosch F, Miksa M, Bittigau P, Vockler J, Dikranian K, Tenkova TI, Stefovska V, Turski L, Olney JW. 1999. Blockade of NMDA receptors and apoptotic neurodegeneration in the developing brain. Science 283(5398):70-74.

Ikonomidou C, Bittigau P, Ishimaru MJ, Wozniak DF, Koch C, Genz K, Price MT, Stefovska V, Horster F, Tenkova T, Dikranian K, Olney JW. 2000. Ethanol-induced apoptotic neurodegeneration and fetal alcohol syndrome. Science 287(5455):1056-1060.

ILAR. 2009. Pain and Distress in Fish. ILAR J 50:327-418.

Jain S, Kumar P, McMillan DD. 2006. Prior leg massage decreases pain responses to heel stick in preterm babies. J Paediatr Child Health 42(9):505-508.

Jevtovic-Todorovic V, Hartman RE, Izumi Y, Benshoff ND, Dikranian K, Zorumski CF, Olney JW, Wozniak DF. 2003. Early exposure to common anesthetic agents causes widespread neurodegeneration in the developing rat brain and persistent learning deficits. J Neurosci 23(3):876-882.

John ER, Prichep LS. 2005. The anesthetic cascade: A theory of how anesthesia suppresses consciousness. Anesthesiology 102(2):447-471.

Johnson JA, Roberston SA, Pypendop BH. 2007. Antinociceptive effects of butorphanol, buprenorphine, or both, administered intramuscularly in cats. Am J Vet Res 68(7): 699-703.

Kamibayashi T, Maze M. 2000. Clinical uses of alpha$_2$-adrenergic agonists. Anesthesiology 93(5):1345-1349.

Karas A, Silverman J. 2006. Pain and distress. In: Suckow M, Silverman J, Murthy S, eds. The IACUC Handbook. Boca Raton, FL: CRC Press.

Karas A, Gostyla K, Aronovitz M, Wolfe E, Karas R. 2001. Diminished body weight and activity patterns in mice following surgery: Implications for control of post-procedural pain/distress in laboratory animals. Contemp Top Lab Anim Sci 40:83.

Karas A, Danneman P, Cadillac J. 2008. Strategies for Assessing and Minimizing Pain in Anesthesia and Analgesia in Laboratory Animals. Fish R, Brown M, Danneman P, Karas A, eds. San Diego: Academic Press. pp. 195-218.

Kehlet H. 2004. Effect of postoperative pain treatment on outcome; Current status and future strategies. Langenbecks Arch Surg 389(4):244-249.

Kehlet H, Dahl JB. 2003. Anaesthesia, surgery, and challenges in postoperative recovery. Lancet 362(9399):1921-1928.

Kehlet H, Jensen TS, Woolf CJ. 2006. Persistent postsurgical pain: Risk factors and prevention. Lancet 367(9522):1618-1625.

Kissin I. 2005. Preemptive analgesia at the crossroad. Anesth Analg 100(3):754-756.

Knotkova H, Pappagallo M. 2007. Adjuvant analgesics. Med Clin North Am 91(1):113-124.

Koblin DD. 2002. Urethane: Help or hindrance? Anesth Analg 94(2):241-242.

Kolesnikov YA, Chereshnev I, Pasternak GW. 2000. Analgesic synergy between topical lidocaine and topical opioids. J Pharmacol Exp Ther 295(2):546-551.

Kona-Boun JJ, Silim A, Troncy E. 2005. Immunologic aspects of veterinary anesthesia and analgesia. JAVMA 226(3):355-363.

Koppert W, Weigand M, Neumann F, Sittl R, Schuettler J, Schmelz M, Hering W. 2004. Perioperative intravenous lidocaine has preventive effects on postoperative pain and morphine consumption after major abdominal surgery. Anesth Analg 98(4):1050-1055.

Kroin JS, Buvanendran A, Watts DE, Saha C, Tuman KJ. 2006. Upregulation of cerebrospinal fluid and peripheral prostaglandin E 2 in a rat postoperative pain model. Anesth Analg 103(2):334-343.

Krugner-Higby L, Smith L, Clark M, Heath TD, Dahly E, Schiffman B, Hubbard-VanStelle S, Ney D, Wendland A. 2003. Liposome-encapsulated oxymorphone hydrochloride provides prolonged relief of postsurgical visceral pain in rats. Comp Med 53(3):270-279.

Krugner-Higby L, Smith L, Heath TD. 2008. Novel delivery systems for analgesic frugs in laboratory animals. In: Fish RE, Brown MJ, Danneman PJ, Karas AZ, eds. Anesthesia and Analgesia in Laboratory Animals, 2nd ed. New York: Elsevier.

Kuan CY, Roth KA, Flavell RA, Rakic P. 2000. Mechanisms of programmed cell death in the developing brain. Trends Neurosci 23(7):291-297.

KuKanich B, Papich MG. 2004. Pharmacokinetics of tramadol and the metabolite O-desmethyltramadol in dogs. J Vet Pharmacol Ther 27(4):239-246.

Kuwahara RT, Skinner RB. 2001. EMLA versus ice as a topical anesthetic. Dermatol Surg 27(5):495-496.

Lagerweij E, Nelis PC, Wiegant VM, van Ree JM. 1984. The twitch in horses: A variant of acupuncture. Science 225(4667):1172-1174.

Lamont L, Mathews K. 2007. Opioids, non-steroidal anti-inflammatories, and analgesic adjuvants. In: Tranquilli WJ, Thurmon JC, Grimm KA, Lumb WV, eds. Lumb and Jones' Veterinary Anesthesia and Analgesia. Ames, IA: Blackwell Publishing.

LASA (Laboratory Animal Science Association). 1990. The assessment and control of the severity of scientific procedures on laboratory animals. Report of the Laboratory Animal Science Association Working Party. Lab Anim 24(2):97-130.

Lascelles BD, Waterman AE, Cripps PJ, Livingston A, Henderson G. 1995. Central sensitization as a result of surgical pain: Investigation of the pre-emptive value of pethidine for ovariohysterectomy in the rat. Pain 62(2):201-212.

Lascelles BD, Cripps PJ, Jones A, Waterman AE. 1997. Post-operative central hypersensitivity and pain: The pre-emptive value of pethidine for ovariohysterectomy. Pain 73(3):461-471.

Lascelles BD, Court MH, Hardie EM, Robertson SA. 2007. Nonsteroidal anti-inflammatory drugs in cats: A review. Vet Anaesth Analg 34(4):228-250.

Laule GE, Bloomsmith MA, Schapiro SJ. 2003. The use of positive reinforcement training techniques to enhance the care, management, and welfare of primates in the laboratory. J Appl Anim Welf Sci 6(3):163-173.

Leece EA, Brearley JC, Harding EF. 2005. Comparison of carprofen and meloxicam for 72 hours following ovariohysterectomy in dogs. Vet Anaesth Analg 32(4):184-192.

Lees P. 2003. Pharmacology of drugs used to treat osteoarthritis in veterinary practice. Inflammopharmacology 11(4):385-399.

Lees P, Giraudel J, Landoni MF, Toutain PL. 2004. PK-PD integration and PK-PD modelling of nonsteroidal anti-inflammatory drugs: Principles and applications in veterinary pharmacology. J Vet Pharmacol Ther 27(6):491-502.

Lemaire L, van der Poll T. 2007. Immunomodulatory effects of general anesthetics. In: Vincent J-L, ed. Intensive Care Medicine: Annual Update 2007. New York: Springer. pp. 208-216.

Lemke KA. 2004. Understanding the pathophysiology of perioperative pain. Can Vet J 45(5): 405-413.

Liles JH, Flecknell PA, Roughan J, Cruz-Madorran I. 1998. Influence of oral buprenorphine, oral naltrexone or morphine on the effects of laparotomy in the rat. Lab Anim 32(2): 149-161.

Lindhardt K, Ravn C, Gizurarson S, Bechgaard E. 2000. Intranasal absorption of buprenorphine: In vivo bioavailability study in sheep. Int J Pharm 205(1-2):159-163.

Linton SJ. 2000. A review of psychological risk factors in back and neck pain. Spine 25(9): 1148-1156.

Liu SS, Richman JM, Thirlby RC, Wu CL. 2006. Efficacy of continuous wound catheters delivering local anesthetic for postoperative analgesia: A quantitative and qualitative systematic review of randomized controlled trials. J Am Coll Surg 203(6):914-932.

Loepke AW, Soriano SG. 2008. An assessment of the effects of general anesthetics on developing brain structure and neurocognitive function. Anesth Analg 106(6):1681-1707.

Loepke AW, Priestley MA, Schultz SE, McCann J, Golden J, Kurth CD. 2002. Desflurane improves neurologic outcome after low-flow cardiopulmonary bypass in newborn pigs. Anesthesiology 97(6):1521-1527.

Lu Y, Vera-Portocarrero LP, Westlund KN. 2003. Intrathecal coadministration of D-APV and morphine is maximally effective in a rat experimental pancreatitis model. Anesthesiology 98(3):734-740.

Luger NM, Sabino MA, Schwei MJ, Mach DB, Pomonis JD, Keyser CP, Rathbun M, Clohisy DR, Honore P, Yaksh TL, Mantyh PW. 2002. Efficacy of systemic morphine suggests a fundamental difference in the mechanisms that generate bone cancer vs inflammatory pain. Pain 99(3):397-406.

Luhmann J, Hurt S, Shootman M, Kennedy R. 2004. A comparison of buffered lidocaine versus ELA-Max before peripheral intravenous catheter insertions in children. Pediatrics 113(3 Pt 1):e217-e220.

Lynch JJ 3rd, Wade CL, Zhong CM, Mikusa JP, Honore P. 2004. Attenuation of mechanical allodynia by clinically utilized drugs in a rat chemotherapy-induced neuropathic pain model. Pain 110(1-2):56-63.

Ma D, Williamson P, Januszewski A, Nogaro MC, Hossain M, Ong LP, Franks NP, Maze M. 2007. Xenon mitigates isoflurane-induced neuronal apoptosis in the developing rodent brain. Anesthesiology 106(4):746-753.

Machin KL. 2001. Fish, amphibian, and reptile analgesia. Vet Clin North Am Exot Anim Pract 4(1):19-33.

Machin KL, Tellier LA, Lair S, Livingston A. 2001. Pharmacodynamics of flunixin and ketoprofen in Mallard ducks (*Anas platyrhynchos*). J Zoo Wild Med 32(2):222-229.

Malavasi LM, Nyman G, Augustsson H, Jacobson M, Jensen-Waern M. 2006. Effects of epidural morphine and transdermal fentanyl analgesia on physiology and behaviour after abdominal surgery in pigs. Lab Anim 40(1):16-27.

Martini L, Lorenzini RN, Cinotti S, Fini M, Giavaresi G, Giardino R. 2000. Evaluation of pain and stress levels of animals used in experimental research. J Surg Res 88(2):114-119.

Martucci C, Panerai AE, Sacerdote P. 2004. Chronic fentanyl or buprenorphine infusion in the mouse: Similar analgesic profile but different effects on immune responses. Pain 110(1-2):385-392.

Mathiesen O, Møiniche S, Dahl JB. 2007. Gabapentin and postoperative pain: A qualitative and quantitative systematic review, with focus on procedure. BMC Anesthesiol 7(7):6.

Mattei P, Rombeau JL. 2006. Review of the pathophysiology and management of postoperative ileus. World J Surg 30(8):1382-1391.

Matthews EA, Dickenson AH. 2002. A combination of gabapentin and morphine mediates enhanced inhibitory effects on dorsal horn neuronal responses in a rat model of neuropathy. Anesthesiology 96(3):633-640.

McAuliffe JJ, Joseph B, Vorhees CV. 2007. Isoflurane-delayed preconditioning reduces immediate mortality and improves striatal function in adult mice after neonatal hypoxia-ischemia. Anesth Analg 104(5):1066-1077, tables of contents.

Mc Geown D, Danbury TC, Waterman-Pearson AE, Kestin SC. 1999. Effect of carprofen on lameness in broiler chickens. Vet Rec 144:668-671.

McMillan FD. 2003. A world of hurts: Is pain special? J Am Vet Med Assoc 223(2):183-186.

Medhurst SJ, Walker K, Bowes M, Kidd BL, Glatt M, Muller M, Hattenberger M, Vaxelaire J, O'Reilly T, Witherspoon G, Winter J, Green J, Urban L. 2002. A rat model of bone cancer pain. Pain 96(1-2):129-140.

Meyer RE, Fish RE. 2005. A review of tribromoethanol anesthesia for production of genetically engineered mice and rats. Lab Anim (NY) 34(10):47-52.

Millecamps M, Etienne M, Jourdan D, Eschalier A, Ardid D. 2004. Decrease in non-selective, non-sustained attention induced by a chronic visceral inflammatory state as a new pain evaluation in rats. Pain 109(3):214-224.

Miranda HF, Puig MM, Prieto JC, Pinardi G. 2006. Synergism between paracetamol and nonsteroidal anti-inflammatory drugs in experimental acute pain. Pain 121(1-2):22-28.

Mogil JS, Smith SB, O'Reilly MK, Plourde G. 2005. Influence of nociception and stress-induced antinociception on genetic variation in isoflurane anesthetic potency among mouse strains. Anesthesiology 103(4):751-758.

Mohan S, Stevens CW. 2006. Systemic and spinal administration of the mu-opioid, remifentanil, produces antinociception in amphibians. Eur J Pharmacol 534(1-3):89-94.

Morley S, Eccleston C, Williams A. 1999. Systematic review and meta-analysis of randomized controlled trials of cognitive behaviour therapy and behaviour therapy for chronic pain in adults, excluding headache. Pain 80(1-2):1-13.

Mosley CA. 2005. Anesthesia and analgesia in reptiles. Semin Avian Exot Pet Medic 14(4): 243-262.

Mosley CA, Dyson D, Smith DA. 2003. Minimum alveolar concentration of isoflurane in green iguanas and the effect of butorphanol on minimum alveolar concetration. J Am Vet Med Assoc 222(11):1559-1564.

Mosley CA, Dyson D, Smith DA. 2004. The cardiovascular dose-response effects of isoflurane alone and combined with butorphanol in the green iguana (*Iguana iguana*). Vet Anaesth Analg 31(1):64-72.

Muir WW, Wiese AJ, March PA. 2003. Effects of morphine, lidocaine, ketamine, and morphine-lidocaine-ketamine drug combination on minimum alveolar concentration in dogs anesthetized with isoflurane. Am J Vet Res 64(9):1155-1160.

Mulcahy DM, Tuomi P, Larsen RS. 2003. Differential mortality of male spectacled eiders (*Somateria fischeri*) and king eiders (*Somateria spectabilis*) subsequent to anesthesia with propofol, bupivacaine, and ketoprofen. J Avian Med Surg 17(3):117-123.

Munro G, Erichsen HK, Mirza NR. 2007. Pharmacological comparison of anticonvulsant drugs in animal models of persistent pain and anxiety. Neuropharmacology 53(5):609-618.

Murrell JC, Hellebrekers LJ. 2005. Medetomidine and dexmedetomidine: A review of cardiovascular effects and antinociceptive properties in the dog. Vet Anaesth Analg 32(3): 117-127.

Murrell JC, Johnson CB. 2006. Neurophysiological techniques to assess pain in animals. J Vet Pharmacol Ther 29(5):325-335.

Naidoo V, Duncan N, Bekker, L, Swan G. 2007. Validating the domestic fowl as a model to investigate the pathophysiology of diclofenac in Gyps vultures. Environ Toxicol Pharmacol 24(3):260-266.

Newby NC, Mendonca PC, Gamperl K, Stevens ED. 2006. Pharmacokinetics of morphine in fish: Winter flounder (*Pseudopleuronectes americanus*) and seawater-acclimated rainbow trout (*Oncorhynchus mykiss*). Comp Biochem Physiol C Toxicol Pharmacol 143(3): 275-283.

Nordgreen J, Horsberg TE, Ranheim B, Chen C. 2007. Somatosensory evoked potentials in the telencephalon of Atlantic salmon (*Salmo salar*) following galvanic stimulation of the tail. J Comp Physiol A 193:1235-1242.

Omote K. 2007. Intravenous lidocaine to treat postoperative pain management: Novel strategy with a long-established drug. Anesthesiology 106(1):5-6.

Ong EL, Lim NL, Koay CK. 2000. Towards a pain-free venipuncture. Anaesthesia 55(3): 260-262.

Otto KA. 2008. EEG power spectrum analysis for monitoring depth of anaesthesia during experimental surgery. Lab Anim 42:45-61.

Padgett DA, Glaser R. 2003. How stress influences the immune response. Trends Immunol 24(8):444-448.

Page GG. 2005. Immunologic effects of opioids in the presence or absence of pain. J Pain Symptom Manage 29(5 Suppl):S25-S31.

Page GG, Blakely WP, Ben-Eliyahu S. 2001. Evidence that postoperative pain is a mediator of the tumor-promoting effects of surgery in rats. Pain 90(1-2):191-199.

Panksepp J. 1980. Brief social isolation, pain responsiveness, and morphine analgesia in young rats. Psychopharmacology 72(1):111-112.

Paul-Murphy JR, Brunson DB, Miletic V. 1999. Analgesic effects of butorphanol and buprenorphine in conscious African grey parrots (*Psittacus erithacus* and *Psittacus erithacus timneh*). Am J Vet Res 60(10):1218-1221.

Peart JN, Gross ER, Gross GJ. 2005. Opioid-induced preconditioning: Recent advances and future perspectives. Vascul Pharmacol 42(5-6):211-218.

Perkins FM, Kehlet H. 2000. Chronic pain as an outcome of surgery: A review of predictive factors. Anesthesiology 93(4):1123-1133.

Ploghaus A, Narain C, Beckmann CF, Clare S, Bantick S, Wise R, Matthews PM, Rawlins JN, Tracey I. 2001. Exacerbation of pain by anxiety is associated with activity in a hippocampal network. J Neurosci 21(24):9896-9903.

Pogatzki-Zahn EM, Zahn PK. 2006. From preemptive to preventive analgesia. Curr Opin Anaesthesiol 19(5):551-555.

Post H, Heusch G. 2002. Ischemic preconditioning: Experimental facts and clinical perspective. Minerva Cardioangiol 50(6):569-605.

Price DD, Mao J, Lu J, Caruso FS, Frenk H, Mayer DJ. 1996. Effects of the combined oral administration of NSAIDs and dextromethorphan on behavioral symptoms indicative of arthritic pain in rats. Pain 68(1):119-127.

Qiu HX, Liu J, Kong H, Liu Y, Mei XG. 2007. Isobolographic analysis of the antinociceptive interactions between ketoprofen and paracetamol. Eur J Pharmacol 557(2-3):141-146.

Read MR. 2004. Evaluation of the use of anesthesia and analgesia in reptiles. JAVMA 227: 547-552.

Reichert JA, Daughters RS, Rivard R, Simone DA. 2001. Peripheral and preemptive opioid antinociception in a mouse visceral pain model. Pain 89(2-3):221-227.

Reilly SC, Quinn JP, Cossins AR, Sneddon LU. 2008a. Novel candidate genes identified in the brain during nociception in common carp. Neurosci Letts 437:135-138.

Reilly SC, Quinn JP, Cossins AR, Sneddon LU. 2008b. Behavioural analysis of a nociceptive event in fish: Comparisons between three species demonstrate specific responses. Appl Anim Behav Sci 114(1-2):248-259.

Rennie AE, Buchanan-Smith HM. 2006. Refinement of the use of non-human primates in scientific research. Part I: The influence of humans. Anim Welf 15(3):203-213.

Reuben SS, Buvanendran A. 2007. Preventing the development of chronic pain after orthopaedic surgery with preventive multimodal analgesic techniques. J Bone Joint Surg Am 89(6):1343-1358.

Richebe P, Rivat C, Laulin JP, Maurette P, Simonnet G. 2005. Ketamine improves the management of exaggerated postoperative pain observed in perioperative fentanyl-treated rats. Anesthesiology 102(2):421-428.

Riess ML, Stowe DF, Warltier DC. 2004. Cardiac pharmacological preconditioning with volatile anesthetics: From bench to bedside? Am J Physiol Heart Circ Physiol 286(5): H1603-H1607.

Robertson SA. 2005. Assessment and management of acute pain in cats. J Vet Emerg Crit Care 15(4):261-272.

Robertson SA, Lascelles BD, Taylor PM, Sear JW. 2005a. PK-PD modeling of buprenorphine in cats: Intravenous and oral transmucosal administration. J Vet Pharmacol Ther 28(5):453-460.

Robertson SA, Taylor PM, Sear JW, Keuhnel G. 2005b. Relationship between plasma concentrations and analgesia after intravenous fentanyl and disposition after other routes of administration in cats. J Vet Pharmacol Ther 28(1):87-93.

Romanovsky AA. 2004. Signaling the brain in the early sickness syndrome: Are sensory nerves involved? Front Biosci 9:494-504.

Romero-Sandoval EA, Horvath RJ, DeLeo JA. 2008. Neuroimmune interactions and pain: Focus on glial-modulating targets. Curr Opin Investig Drugs 9:726-734.

Ross JR, Riley J, Quigley C, Welsh KI. 2006. Clinical pharmacology and pharmacotherapy of opioid switching in cancer patients. Oncologist 11(7):765-773.

Roughan JV, Flecknell PA. 2001. Behavioural effects of laparotomy and analgesic effects of ketoprofen and carprofen in rats. Pain 90(1-2):65-74.

Roughan JV, Flecknell PA. 2004. Behaviour-based assessment of the duration of laparotomy-induced abdominal pain and the analgesic effects of carprofen and buprenorphine in rats. Behav Pharmacol 15(7):461-472.

Roy S, Wang J, Charboneau R, Loh HH, Barke RA. 2005. Morphine induces CD4+ T cell IL-4 expression through an adenylyl cyclase mechanism independent of the protein kinase A pathway. J Immunol 175(10)6361-6367.

Sacerdote P, Bianchi M, Gaspani L, Manfredi B, Maucione A, Terno G, Ammatuna M, Panerai AE. 2000. The effects of tramadol and morphine on immune responses and pain after surgery in cancer patients. Anesth Analg 90(6):1411-1414.

Sakai H, Sheng H, Yates RB, Ishida K, Pearlstein RD, Warner DS. 2007. Isoflurane provides long-term protection against focal cerebral ischemia in the rat. Anesthesiology 106(1):92-99.

Samad TA, Moore KA, Sapirstein A, Billet S, Allchorne A, Poole S, Bonventre JV, Woolf CJ. 2001. Interleukin-1 beta-mediated induction of Cox-2 in the CNS contributes to inflammatory pain hypersensitivity. Nature 410:471-475.

Samad TA, Sapirstein A, Woolf CJ. 2002. Prostanoids and pain: Unraveling mechanisms and revealing therapeutic targets. Trends Mol Med 8(8):390-396.

Sasamura T, Nakamura S, Iida Y, Fujii H, Murata J, Saiki I, Nojima H, Kuraishi Y. 2002. Morphine analgesia suppresses tumor growth and metastasis in a mouse model of cancer pain produced by orthotopic tumor inoculation. Eur J Pharmacol 441(3):185-191.

Schipke JD, Kerendi F, Gams E, Vinten-Johansen J. 2006. Postconditioning: A brief review. Herz 31(6):600-606.

Schneemilch CE, Hachenberg T, Ansorge S, Ittenson A, Bank U. 2005. Effects of different anaesthetic agents on immune cell function in vitro. Eur J Anaesthesiol 22(8):616-623.

Shäfers M, Marziniak M, Sorkin LS, Yaksh TL, Sommer C. 2004. Cyclo-oxygenase inhibition in nerve injury- and TNF-induced hyperalgesia in the rat. Exp Neurol 185(1):160-168.

Shaiova L. 2006. Difficult pain syndromes: Bone pain, visceral pain, and neuropathic pain. Cancer J 12(5):330-340.

Shavit Y, Fish G, Wolf G, Mayburd E, Meerson Y, Yirmiya R, Beilin B. 2005. The effects of perioperative pain management techniques on food consumption and body weight after laparotomy in rats. Anesth Analg 101(4):1112-1116.

Sladky KK, Miletic V, Paul-Murphy J, Kinney ME, Dallwig RK, Johnson SM. 2007. Analgesic efficacy and respiratory effects of butorphanol and morphine in turtles. JAVMA 230(9):1356-1362.

Slikker W Jr, Zou X, Hotchkiss CE, Divine RL, Sadovova N, Twaddle NC, Doerge DR, Scallet AC, Patterson TA, Hanig JP, Paule MG, Wang C. 2007. Ketamine-induced neuronal cell death in the perinatal rhesus monkey. Toxicol Sci 98(1):145-158.

Smith ST. 2007. Compendium of drugs and compounds used in amphibians. ILAR J 48(3): 297-300.

Smith LJ, Krugner-Higby L, Trepanier LA, Flaska DE, Joers V, Heath TD. 2004. Sedative effects and serum drug concentrations of oxymorphone and metabolites after subcutaneous administration of a liposome-encapsulated formulation in dogs. J Vet Pharmacol Ther 27(5):369-372.

Sneddon LU. 2003. The evidence for pain in fish: The use of morphine as an analgesic. Appl Anim Behav Sci 83(2):153-162.

Sneddon LU. 2004. Evolution of nociception in vertebrates: Comparative analysis of lower vertebrates. Brain Res Brain Res Rev 46(2):123-130.

Sneddon LU. 2006. Ethics and welfare: Pain perception in fish. B Eur Assoc Fish Path 26(1): 6-10.

Sneddon LU, Braithwaite VA, Gentle MJ. 2003a. Do fishes have nociceptors? Evidence for the evolution of a vertebrate sensory system. Proc R Soc Lond B Biol Sci 270(1520): 1115-1121.

Sneddon LU, Braithwaite VA, Gentle MJ. 2003b. Novel object test: Examining nociception and fear in the rainbow trout. Pain 4(8):431-440.

Soriano SG, Anand KJS, Rovnaghi CR, Hickey PR. 2005. Of mice and men: Should we extrapolate rodent experimental data to the care of human neomates? Anesthesiology 102:866-868.

Stamer UM, Lehnen K, Hothker F, Bayerer B, Wolf S, Hoeft A, Stuber F. 2003. Impact of CYP2D6 genotype on postoperative tramadol analgesia. Pain 105(1-2):231-238.

Stefano GB, Fricchione G, Goumon Y, Esch T. 2005. Pain, immunity, opiate and opioid compounds and health. Med Sci Monit 11(5):MS47-MS53.

Stevens CW. 2004. Opioid research in amphibians: An alternative pain model yielding insights on the evolution of opioid receptors. Brain Res Brain Res Rev 46(2):204-215.

Stevens CW, Klopp AJ, Facello JA. 1994. Analgesic potency of mu and kappa opioids after systemic administration in amphibians. J Pharmacol Exp Ther 269(3):1086-1093.

Stevens CW, MacIver DN, Newman LC. 2001. Testing and comparison of non-opioid analgesics in amphibians. Contemp Top Lab Anim 40(4):23-27.

Stoskopf MK. 1994. Pain and analgesia in birds, reptiles, amphibians, and fish. Invest Ophthalmol Vis Sci 35(2):775-780.

Strigo IA, Duncan GH, Bushnell MC, Boivin M, Wainer I, Rodriguez Rosas ME, Persson J. 2005. The effects of racemic ketamine on painful stimulation of skin and viscera in human subjects. Pain 113(3):255-264.

Suleiman MS, Zacharowski K, Angelini GD. 2008. Inflammatory response and cardioprotection during open-heart surgery: The importance of anaesthetics. Br J Pharmacol 153(1): 21-33.

Tariot PN, Farlow MR, Grossberg GT, Graham SM, McDonald S, Gergel I, the Memantine Study Group. 2004. Memantine treatment in patients with moderate to severe Alzheimer disease already receiving donepezil: A randomized controlled trial. JAMA 291:317-324.

Terner JM, Lomas LM, Smith ES, Barrett AC, Picker MJ. 2003. Pharmacogenetic analysis of sex differences in opioid antinociception in rats. Pain 106(3):381-391.

Terril-Robb LA, Suckow MA, Grigdesby CF. 1996. Evaluation of the analgesic effects of butorphanol tartrate, xylazine hydrochloride, and flunixin meglumine in Leopard Frogs (*Rana pipiens*). Contemp Top Lab Anim 35(3):54-56.

Todd MM. 2004. Anesthetic neurotoxicity: The collision between laboratory neuroscience and clinical medicine. Anesthesiology 101(2):272-273.

Tranquilli WJ, Thurmon JC, Grimm KA, Lumb WV, eds. 2007. Lumb and Jones' Veterinary Anesthesia and Analgesia. Ames, Iowa: Blackwell Publishing.

Trnkova S, Knotkova Z, Hrda A, Knotek Z. 2007. Effect of non-steroidal anti-inflammatory drugs on the blood profile in the green iguana (*Iguana iguana*). Veter Medic 52:507-511.

Tuttle AD, Papich M, Lewbart GA, Christian S, Gunkel C, Harms CA. 2006. Pharmacokinetics of ketoprofen in the green iguana (*Iguana iguana*) following single intravenous and intramuscular injections. J Zoo Wildl Med 37(4):567-570.

Ulrich-Lai YM, Xie W, Meij JT, Dolgas CM, Yu L, Herman JP. 2006. Limbic and HPA axis function in an animal model of chronic neuropathic pain. Physiol Behav 88(1-2):67-76.

Vallejo R, Hord ED, Barna SA, Santiago-Palma J, Ahmed S. 2003. Perioperative immunosuppression in cancer patients. J Environ Pathol Toxicol Oncol 22(2):139-146.

Valverde A, Gunkel CI. 2005. Pain management in horses and farm animals. J Vet Emerg Crit Car 15(4):295-307.

Visser E, Schug SA. 2006. The role of ketamine in pain management. Biomed Pharmacother 60:341-348.

Wagner KA, Gibbon KJ, Strom TL, Kurian JR, Trepanier LA. 2006. Adverse effects of EMLA (lidocaine/prilocaine) cream and efficacy for the placement of jugular catheters in hospitalized cats. J Feline Med Surg 8(2):141-144.

Waldhoer M, Bartlett SE, Whistler JL. 2004. Opioid receptors. Annu Rev Biochem 73: 953-990.

Wang L, Traystman RJ, Murphy SJ. 2008. Inhalational anesthetics as preconditioning agents in ischemic brain. Curr Opin Pharmacol 8(1):104-110.

Waterman AE, Livingston A, Amin A. 1991. Further studies on the antinociceptive activity and respiratory effects of buprenorphine in sheep. J Vet Pharmacol Ther 14(3):230-234.

Watkins LR, Maier SF. 2005. Immune regulation of central nervous system functions: From sickness responses to pathological pain. J Intern Med 257(2):139-155.

Weber NC, Preckel B, Schlack W. 2005. The effect of anaesthetics on the myocardium: New insights into myocardial protection. Eur J Anaesthesiol 22(9):647-657.

Weise KL, Nahata MC. 2005. EMLA for painful procedures in infants. J Pediatr Health Care 19(1):42-47; quiz 48-49.

Whelan G, Flecknell PA. 1992. The assessment of depth of anaesthesia in animals and man. Lab Anim 26(3):153-162.

White PF. 2005. The changing role of non-opioid analgesic techniques in the management of postoperative pain. Anesth Analg 101(5 Suppl):S5-S22.

White PF, Kehlet H, Neal JM, Schricker T, Carr DB, Carli F. 2007. The role of the anesthesiologist in fast-track surgery: From multimodal analgesia to perioperative medical care. Anesth Analg 104(6):1380-1396.

Whiteside GT, Harrison J, Boulet J, Mark L, Pearson M, Gottshall S, Walker K. 2004. Pharmacological characterisation of a rat model of incisional pain. Br J Pharmacol 141(1):85-91.

Wieseler-Frank J, Maier SF, Watkins LR. 2005. Immune-to-brain communication dynamically modulates pain: Physiological and pathological consequences. Brain Behav Immun 19(2):104-111.

Willenbring S, Stevens CW. 1997. Spinal mu, delta and kappa opioids alter chemical, mechanical and thermal sensitivities in amphibians. Life Sci 61(22):2167-2176.

Wilson SG, Bryant CD, Lariviere WR, Olsen MS, Giles BE, Chesler EJ, Mogil JS. 2003a. The heritability of antinociception II: Pharmacogenetic mediation of three over-the-counter analgesics in mice. J Pharmacol Exp Ther 305(2):755-764.

Wilson SG, Smith SB, Chesler EJ, Melton KA, Haas JJ, Mitton B, Strasburg K, Hubert L, Rodriguez-Zas SL, Mogil JS. 2003b. The heritability of antinociception: Common pharmacogenetic mediation of five neurochemically distinct analgesics. J Pharmacol Exp Ther 304(2):547-559.

Wolfensohn S. 2004. Social housing of large primates: Methodology for refinement of husbandry and management. ATLA Altern Lab Anim 32(Suppl 1A):149-151.

Woolf CJ. 1983. Evidence for a central component of postinjury pain hypersensitivity. Nature 308:386-388.

Woolf CJ. 2007. Central sensitization: Uncovering the relation between pain and plasticity. Anesthesiology 106:864-867.

Wright-Williams SL, Courade JP, Richardson CA, Roughan JV, Flecknell PA. 2007. Effects of vasectomy surgery and meloxicam treatment on faecal corticosterone levels and behaviour in two strains of laboratory mouse. Pain 130(1-2):108-118.

Xie Z, Dong Y, Maeda U, Moir RD, Xia W, Culley DJ, Crosby G, Tanzi RE. 2007. The inhalation anesthetic isoflurane induces a vicious cycle of apoptosis and amyloid beta-protein accumulation. J Neurosci 27(6):1247-1254.

Yaksh TL, Dirig DM, Malmberg AB. 1998. Mechanism of action on non-steroidal anti-inflammatory drugs. Cancer Investig 16(7):509-527.

Yamamoto K, Ngan MP, Takeda N, Yamtodani A, Rudd JA. 2004. Differential activity of drugs to induce emesis and pica behavior in *Suncus murinus* (house musk shrew) and rats. Physiol Behav 83:151-156.

Yoon WY, Chung SP, Lee HS, Park YS. 2008. Analgesic pretreatment for antibiotic skin test: Vapocoolant spray vs ice cube. Am J Emerg Med 26(1):59-61.

Zhao P, Zuo Z. 2004. Isoflurane preconditioning induces neuroprotection that is inducible nitric oxide synthase-dependent in neonatal rats. Anesthesiology 101(3):695-703.

5

Humane Endpoints for Animals in Pain

This chapter presents an overview of the concept of humane endpoints and their application in studies that cause pain in research animals. It sets the stage with a review of pertinent guidance documents, focusing on the Organization for Economic Cooperation and Development (OECD) 2000 Guidance on Humane Endpoints for Experimental Animals Used in Safety Evaluation. It provides a discussion of the usefulness of pilot studies as a refinement and potential replacement tool. Further, it presents humane endpoints in relation to specific research fields—toxicology, infectious diseases, vaccine safety, cancer, and pain. It concludes with a discussion of euthanasia.

GUIDELINES AND REFERENCE DOCUMENTS

Moral and ethical obligations are inherent in all aspects of research, testing, and teaching that use research subjects. The question of when a study using animal models should end or the study design be changed due to animal pain, distress, or welfare considerations has been the subject of many publications, symposia, guidance documents, and regulations. Defining a humane endpoint can vary widely depending on a number of factors, of which study design and research objectives are but two. Consequently, attempting to provide specific endpoint criteria for all study designs and other factors cannot be adequately addressed in this one report (Morton 1999, 2000). Not only would such a list be inadequate, it could prove detrimental to hitherto unknown study objectives. This report does not go into specifics but rather presents selected pertinent guidelines and documents.

Investigators, study personnel, veterinary staff, and institutional animal care and use committees (IACUCs) are obligated to thoroughly research and incorporate humane endpoints in every study or use involving laboratory animals.

National and International Guidelines

A number of national and international guidelines are available to assist researchers in determining humane endpoints for research animals. The Office of Laboratory Animal Welfare (OLAW) defines these as "[c]riteria used to end experimental studies earlier in order to avoid or terminate unrelieved pain and/or distress are referred to as humane endpoints. An important feature of humane endpoints is that they should ensure that study objectives will still be met even though the study is ended at an earlier point. Ideally, humane endpoints are sought that can be used to end studies before the onset of pain and distress" (OLAW/ARENA 2002, p. 103).

The Canadian Council for Animal Care (CCAC) has published an excellent document with general recommendations on humane endpoints in animal studies. According to the CCAC guidelines, in "experiments involving animals, any actual or potential pain, distress, or discomfort should be minimized or alleviated by choosing the earliest endpoint that is compatible with the scientific objectives of the research. Selection of this endpoint by the investigator should involve consultation with the laboratory animal veterinarian and the animal care committee" (CCAC 1998, p. 5).

In 1994, the OECD recognized that while ambiguous test guidelines may be necessary, such ambiguity fosters an overbroad interpretation of what constitutes a humane endpoint in toxicology studies. The organization therefore created a working group to develop a guidance document using clinical signs as humane endpoints in safety evaluation studies (OECD 2000; Box 5-1). The resulting document put forth criteria based on the principles of the 3Rs as well as descriptions of clinical signs to assist study personnel in determining when death may be imminent or when severe pain may be present after an animal's exposure to a test substance. The criteria are broad enough to apply to a wide range of study types, test substances, species, and strains of animals. The reader is encouraged to examine this resource when developing internal guidance documents to assess humane endpoints.

OECD invested considerable time and effort in addressing and defining potential endpoints in safety assessment studies (see the Addendum at the end of this chapter for the OECD definition). The OECD Guidance Document defines humane endpoints "as the earliest indicator in an animal experiment of severe pain, severe distress, suffering, or impending death. The ultimate purpose of the application of humane endpoints to toxicol-

BOX 5-1
OECD Guidance Document on the Recognition, Assessment, and Use of Clinical Signs as Humane Endpoints for Experimental Animals Used in Safety Evaluation (OECD 2000)

- A humane endpoint can be defined as the earliest indicator in an animal experiment of severe pain, severe distress, suffering, or impending death.
- The ultimate purpose of the application of humane endpoints to toxicology studies is to be able to accurately predict severe pain, severe distress, suffering, or impending death, before the animal experiences these effects. However, the science of toxicology is not yet to the point where such accurate predictions can be made *prior to* the onset of severe pain and distress. It is possible at this time to identify pain, distress, or suffering, very early after their onset by careful clinical examination of animals on test using well-defined endpoints and criteria. Humane endpoints for use in research and testing have been addressed in a number of publications. . . . These adverse conditions, once identified, should be minimized or eliminated, either by humanely killing the animal or, in long-term studies, by (temporary) termination of exposure, or by reduction of the test substance dose.
- Different animal species, and animals at different stages of development, may respond differently to test conditions, and exhibit different indications of distress. The clinical signs described here should be evaluated in consideration of these potential differences. If relevant humane endpoints have been identified, they should be described when an experiment is being planned, and incorporated into the experimental protocol and all related standard operating procedures (SOPs).

ogy studies is to be able to accurately predict severe pain, severe distress, suffering, or impending death, before the animal experiences these effects" (OECD 2000, p. 10). While the OECD indicated that the science of toxicology cannot accurately predict pain prior to onset, careful observations can "identify pain, distress, or suffering, very early after their onset . . . using well-defined endpoints and criteria." The OECD further advises that suffering "should be minimized or eliminated, either by humanely killing the animal or, in long-term studies, by (temporary) termination of exposure, or by reduction of the test substance dose. Different animal species, and animals at different stages of development, may respond differently to test conditions, and exhibit different indications of distress" (ibid.).

These guidance documents are consistent in their recommendations. Predictive parameters must be reliable, reproducible, and objective, and allow both the achievement of study objectives and goals and the use of appropriate methodologies at the earliest point to alleviate or avoid pain. As discussed below, pilot studies are an effective means to identify and validate humane endpoints, which can then be incorporated in research methods to minimize, alleviate, or avoid pain for the animal subjects (also see Morton 1999, 2000; Stokes 2002; NRC 2008, p. 61).

Humane endpoints were the focus of a 1998 international conference in Ziest, The Netherlands. The editors of the conference proceedings determined that humane endpoints are specific to individual studies or a particular testing paradigm (Hendriksen and Morton 1999, pp. v-vi), based on study design and intent, regulatory requirements, personnel connected to the study, and the animals themselves, whether as individuals or as a group. The conference participants concluded that the establishment of humane endpoints is, and should be, subject to adaptation as societal mores, attitudes, regulations, and technologies change. The conference report further stated that for ethical reasons, the formulation of endpoints to avoid or alleviate pain in laboratory animals must be a high ethical priority in every facility that conducts any form of animal experimentation (ibid.).

Beyond Formal Guidelines

Many of the articles and recommendations that address humane endpoints focus on very specific study or research types that can cause pain to laboratory animals; for example, studies on the identification and use of humane endpoints in animal models of sepsis and shock provide an excellent overview of the methodologies to determine humane endpoints yet still achieve study objectives (Nemzek et al. 2004, 2008). More generally, the Institute for Laboratory Animal Research (ILAR) bases its reports on its mission statement promoting "high-quality science and humane care and use of research animals based on the principles of refinement, replacement, and reduction (the 3Rs) and high ethical standards" (ILAR 2009). The Institute's *Guidelines for the Care and Use of Mammals in Neuroscience and Behavioral Research* (NRC 2003) provide criteria for evaluating levels of pain that help in the development of endpoints for studies in neuroscience and behavioral research. An *ILAR Journal* issue dedicated to Humane Endpoints for Animals Used in Biomedical Research and Testing (ILAR 2000) provides an overview of several research areas where pain is a potential outcome, including infectious disease and cancer research (Olfert and Godson 2000; Wallace 2000) and vaccine potency and and acute toxicity testing (Hendriksen and Steen 2000; Sass 2000). ILAR also published the proceedings of a symposium on Regulatory Testing and Animal Welfare, detailing best practices for the humane conduct of animal testing for regulatory purposes (NRC 2004).

While these references are extremely valuable, it is important to view them in accordance with their intent: they are guidance documents only and as such have limitations. No single document could cover all potentially painful study types, all animal species used in research, or all clinical signs associated with all research projects. In the absence of comprehensive guidance, the scientific community has an ethical responsibility to develop a general humane endpoint policy at each institution to provide guidance and a basis for dialogue between scientists and IACUCs about individual protocols.

Caution is advisable, however, in efforts to develop a policy on humane endpoints. While the ideal is to avoid pain, personnel also need to ensure that the study objectives are attained before a procedure or animal is terminated (OLAW/ARENA 2002, p. 103). If a full study, or aspect of a study, is ended before the objectives have been met, one can argue that the animals used have been wasted. Moreover, if the purpose of a study is to meet the requirements for the safety assessment of a substance, a regulatory agency may reject the submitted data as insufficient and require that the study be repeated. On the other hand, if researchers are reluctant to intervene, study animals may unnecessarily experience pain, distress, or severely diminished welfare. Further, without adequate guidance, death is likely to be selected as a convenient endpoint that is reproducible and objective. If regulatory guidelines do not specify an endpoint, as in vaccine potency studies (CFR Title 9, 2006), regulated entities can and will use lethality.

For all these reasons identification of humane endpoints should take into account the following factors: the role of regulatory agencies in the overall process; the need for scientifically appropriate endpoints; and the reliability of clinical observations of the animals to ensure a proper outcome for both the animals and the study. As a corollary, it is worth emphasizing that investigators, technicians, and other staff responsible for the care of research animals should be well trained and able to make impartial judgments about an animal's well-being.

OLAW approached the subject of humane endpoints in its *Institutional Animal Care and Use Committee Guidebook* (OLAW/ARENA 2002, p. 103), advising internal oversight committees to review protocols to determine whether "discomfort to animals will be limited to that which is unavoidable for the conduct of scientifically valuable research, and [whether] unrelieved pain and distress will only continue for the duration necessary to accomplish the scientific objectives." The OLAW reference is careful to state that potential pain or distress should be relieved with appropriate medication or with euthanasia, although the study objectives should still be met. The intent is to end a study before the development of pain or distress, as is emphasized in the OECD document.

PILOT STUDIES

An effective way to reduce negative impacts on laboratory animals is the use of a pilot study, which can be critical to the success of a larger study (DeHaven 2002; Morton et al. 1990; NRC 2003, p. 14; NRC 2008, pp. 61-62; OECD 2000, p. 14). The premise behind this concept is to conduct the proposed study on a small number of animals rather than the full complement necessary for a statistically valid study and thus prevent unnecessary pain for a larger number of animals.

Pilot studies are advantageous because they help researchers to identify:

- potential interactions between proposed analgesic and anesthetic treatments and specific research goals,
- potentially useful means of assessing pain in a specific research model, and
- humane endpoint criteria specific to an individual project.

Problems that occur in the pilot study can inform the discussion and development of strategies to address an animal's deteriorating condition. Such strategies may include (but certainly not be limited to) the adjustment of dose levels, changes in sample size, identification of adverse effects, incorporation of refinements (e.g., use of analgesics, procedural changes), or alteration to the duration of exposure to minimize negative impacts on the animals.

Caution is essential in the design and conduct of pilot studies as the risk of causing significant pain to the animals in such studies can be high. This risk necessitates close oversight by the IACUC and careful monitoring of the animals by study personnel and veterinary staff. Good communication among all involved can ensure both the collection of the maximum amount of useful data and appropriate interventions on behalf of the animals (NRC 2003, p. 14).

INTERNATIONAL REGULATIONS AND GUIDELINES FOR SAFETY ASSESSMENT

Regulatory bodies in most countries have developed standards and guidelines to ensure the conduct of appropriate safety assessments on test substances (Hicks 1997; Merrill 2001; USEPA 2008). For example, after the use of thalidomide by pregnant women in the 1960s caused severe birth defects in the long bones of the fetuses, US legislation required adequate testing of drugs in animals before human exposure (Gallo 2001; Nies 2001).

Similar legislative actions followed environmental disasters like the Love Canal contamination (Merrill 2001).

The purpose of testing requirements for pharmaceutical, consumer, and industrial products is to ensure the safety of the environment and of the human and animal populations. However, these requirements have tended to focus on the safety of the user and do not necessarily consider humane endpoints for the animals used in the safety assessment, although such consideration is becoming a more prominent component of some newer regulatory requirements.

In June 2007, the European Commission established a regulation to evaluate the hazards and risks of chemicals (Regulation (EC) No. 1907/2006 of the European Parliament and of the Council of 18 December 2006); the mission of REACH (Registration, Evaluation, and Authorization of Chemicals) is to improve the assessment of chemicals in order to better protect human health and the environment. Because the range of chemicals covered by REACH is enormous, there is great potential for increased use of animals in corresponding toxicity and safety testing. But the regulation ensures the authorization of animal testing only when necessitated by identification of data gaps (ECHA 2008). Furthermore, the regulation requires industry to share data on similar chemicals to avoid duplicative animal testing; allows for the submission of data using nonanimal tests; strongly encourages the use of Quantitative Structure-Activity Relationship (QSAR) or other computer-generated information; and invites the grouping of submitted data for similar chemicals that may result in similar hazards and risks (the so-called "read-across" principle). While these efforts do not define humane endpoints, the authors of the regulation are commended for the consideration of responsible animal use in safety assessment.

Also useful in the toxicology regulatory arena is a February 2008 Memorandum of Understanding (MOU) that lays the foundation and framework for the US Environmental Protection Agency (EPA) and two NIH agencies to collaborate in sharing data, resources, and expertise in efforts to replace animal testing for chemical toxicity assessment (Collins et al. 2008; NIH/USEPA 2008; NIH 2008). The MOU calls for the evaluation of in vitro assays, such as those used for identification of toxicity pathways and high-throughput screening (as described in NRC 2007), to better predict potential health and environmental hazards from chemicals. The ambitious goals of the MOU are the development of more accurate assays and changes in regulatory guidelines, both of which are likely to be a long-term process. Similar goals should be encouraged on a global scale to effect change in regulatory agencies and eliminate potentially painful animal testing.

Although harmonization of regulatory guidelines has significantly reduced discrepancies between cooperating countries, efforts for the global harmonization of safety guidelines are neither consistent nor well coordinated. As a result, tests must comply with all the requirements of each country where a product is to be marketed for a particular use. For example, the regulatory agency of one country may require an additional group of animals to assess recovery from exposure, while other countries may not have this requirement or may even reject the study depending on their review process. Or one country's regulatory agency may accept an alternative that has been validated as scientifically reliable and relevant (NIH 1997), such as the local lymph node assay in mice, whereas agencies in other countries may not accept the data in lieu of the guinea pig dermal sensitization test.

While a comparison of all safety assessment guidelines is well beyond the scope of this report, differences in regulatory-driven studies can have a negative impact on the prevention and alleviation of pain in laboratory animals. An example of a safety assessment test that may cause pain is the acute eye irritation study, the purpose of which is to evaluate the potential hazards of ocular exposure to a substance. Although requirements for this procedure are generally in agreement across international regulatory bodies and national agencies (JMAFF 2000; OECD 1987, #405; USEPA OPPTS 1998, #870.2400), the same is not true for the reversibility of ocular lesions, an additional requirement of this test in order to more fully assess the risk of human exposure. The procedures for this component of the toxicity evaluation vary considerably with respect to animal welfare. The OECD guidelines recommend a step-wise evaluation paradigm that starts with assessment of structurally related substances and other in vitro tests prior to any animal use. The guidelines also identify ocular lesions that are considered irreversible and thereby meet OECD criteria for terminating the study and euthanizing the animal. But while guidelines in various countries reference the OECD guidance document for humane endpoints and recommend the use of local anesthetics in cases of extreme pain, they do not recognize the OECD criterion for early termination of the study (identification of irreversible lesions).

HUMANE ENDPOINTS IN TOXICOLOGY STUDIES

In recognition of the pain and distress inflicted on animals in many safety and toxicology studies, regulatory guidelines have begun to address the concept of humane endpoints, although sometimes in vague terms. The EPA Health Effects Test Guidelines for Acute Oral Toxicity (USEPA OPPTS 2002) provide instruction for following the OECD Guidance Document (OECD 2000) to reduce the suffering of animals in toxicity studies. Euthanasia of animals that are either moribund or in severe pain is also encouraged. Regrettably, vague statements such as "animals showing severe and enduring signs of distress and pain may need to be humanely killed," which are common in regulatory guidelines (USEPA OPPTS 1998), may promote a reluctance to terminate a study or an animal's exposure to the testing substance because a regulatory body may consider the action premature and mandate a repeat study. This is not a good situation for researchers, laboratories, or animals.

Not all test substances cause ocular (or other) pain or injury, but the potential exists. As pointed out by Durham and colleagues (1992), there is a gap in the data for analgesia appropriate for use in ocular toxicity tests and that gap persists, as evidenced in a US Federal Register notice (Federal Register 2007) requesting data on analgesic use in ocular irritancy tests to alleviate pain without affecting test results. Current guidelines include neither justification for withholding analgesic agents nor guidance for the use of analgesic agents to alleviate ongoing pain. As a result, testing entities may be reluctant to provide analgesia beyond initial local anesthetics, to avoid the possibility of interference with the test substance (Stokes 2005). Yet numerous published studies demonstrate that the use of analgesics to alleviate pain from ocular irritancy tests does not interfere with the scientific objectives of this safety test (Patrone et al. 1999; Peyman et al. 1994; Stiles et al. 2003). Such evidence can be used to avoid or alleviate pain as well as to provide scientific rationale for the use of analgesics in ocular irritancy tests.

Chronic toxicity and carcinogenicity testing are currently required to assess effects after long-term, repeated exposure to a test substance (JMAFF 2000; OECD 1987, #405; USEPA OPPTS 1998, #870.2400). The incidence of tumor burden, geriatric changes, and premature death can be significant near the scheduled termination of these studies. Guidelines generally specify the survival rates necessary to provide meaningful interpretation of a chronic study, but the OECD document is the only one to discuss humane endpoints and provide guidance for the early termination of a study if survival rates fall below a specified percentage. In order to achieve the required survival rate at the end of the mandated study, animals often are not euthanized until very close to death, an outcome that may entail needless pain for the animals. True harmonization of guideline safety assessment tests and global adoption of the OECD humane endpoints document would be an important step toward the alleviation and avoidance of pain in laboratory animals.

The NRC report *Toxicity Testing in the Twenty-first Century: A Vision and a Strategy* (NRC 2007) evaluated current toxicity testing schemes and developed a long-term strategy for the direction of safety assessments based on state-of-the-art sciences (e.g., genomics, proteomics, and pharmacokinetics) and emerging technologies (e.g., bioinformatics). Although the report acknowledges that implementation of the strategy will require much effort on the part of scientists, regulators, and law makers to develop workable testing schemes, the concepts envisioned could significantly improve the science of toxicology, assessment of risk to human safety, alleviation of pain in laboratory animals, and reduction or replacement of animals in toxicity testing (ibid.).

One of the sources reviewed for the NRC report was the approach developed by the Health and Environmental Sciences Institute (HESI) of the International Life Sciences Institute (ILSI). In 2000, this organization convened an Agricultural Chemical Safety Assessment (ACSA) committee to redesign safety testing schemes for agricultural chemicals. The resulting multifaceted approach redesigns traditional toxicology tests to integrate several sciences, such as metabolism/kinetics and life stages, in a single study to eliminate the requirement for separate studies to evaluate each parameter and reduce the number of animals used (Carmichael et al. 2006; Cooper et al. 2006; Doe et al. 2006; ILSI-HESI 2008). Further, the metabolism/kinetics component of the strategy is particularly relevant to the alleviation of pain in laboratory animals: based on the metabolism of a test substance in the animal model, a saturation point can be determined and used as the high dose level in subsequent studies because it is considered more relevant to actual human exposure levels. This approach, based on step-wise, or tiered, testing, is expected to reduce animal numbers, minimize potential pain to laboratory animals by avoiding exposure levels that produce clinical signs of toxicity, and improve the quality of data for assessments of risk to humans (Carmichael et al. 2006).

HUMANE ENDPOINTS IN INFECTIOUS DISEASE RESEARCH

There has been an increase in infectious disease research as a result of bioterrorism threats and anthrax attacks since September 11, 2001 (Copps 2005; Jaax 2005). Whether the disease agent is of interest for bioterrorism or for human or animal welfare, the study of a targeted disease typically involves exposing healthy research animals to a disease agent that culminates in clinical disease and death. The animals may experience significant pain during these experiments, but identification and validation of earlier endpoints to safeguard animal welfare can be difficult, as an inappropriate endpoint may not adequately identify the full course of a disease or the efficacy of a potential medication (Olfert and Godson 2000). It is imperative, therefore, to examine and validate endpoints within a solid scientific framework that includes, among others, immunological parameters, biochemical and endocrine changes, and other pathophysiologic changes (e.g., decreased body temperature). Moreover, eliminating death as the endpoint for infectious disease research can benefit not only the laboratory animals but the research itself because pathological changes are easier to identify in fresh tissues as opposed to autolyzed tissues from animals that have been allowed to die (Copps 2005).

HUMANE ENDPOINTS IN VACCINE SAFETY AND POTENCY TESTING

Another area of research that frequently results in the death of study animals is vaccine testing for regulatory agencies. Because vaccines are biological products and one batch may not be as potent as the next or may contain harmful byproducts, it is important to test both their efficacy and safety (Castle 1999; Cussler et al. 1999; Hendriksen 2002). To ensure quality control and the safety of each batch, regulatory agencies such as the US Food and Drug Administration (FDA), the US Department of Agriculture (USDA), the European Pharmacopoeia, and the World Health Organization (WHO) require potency testing during which animals are vaccinated and then exposed to the virulent disease agent. However, the endpoint for each potency test is not consistent across disease agents. In some instances, regulations require that a certain percentage of control animals die before a test is considered valid, while other tests are based on the survival of the vaccinated animals. For example, the FDA-administered safety test for general biological products requires vaccination of healthy guinea pigs and mice with a small dose of the final product from each vaccine lot (CFR 2008, 610.11). A safety test is considered unsatisfactory if the animals do not survive the 7-day test period, in which case additional safety tests over a larger test population are required. The USDA-mandated potency testing for *Leptospira pomona* bacterin (CFR 2006, 113.101) requires that at least eight of ten unvaccinated control animals die in order to validate the test. Other potency testing may require a comparison of death rates in the vaccinated versus control animals, as, for example, in the USDA safety test for Marek's disease vaccine (CFR 2006, 113.330). For this type of testing a more humane endpoint would be the onset of clinical signs in unvaccinated controls; thus for example the potency test for tetanus antitoxin is met when unvaccinated control guinea pigs are unable to stand within 24 hours postchallenge, at which point the animals may be euthanized (CFR 2006, 113.451).

Regulations may also encourage the use of in vitro methods. The USDA canine distemper killed virus vaccine potency test (CFR 2006, 113.201) accepts serum titer levels in vaccinated animals for potency data; if, however, the tests are inconclusive, a viral challenge test is required, using both vaccinated and unvaccinated controls. The agency identifies the survival of all vaccinated animals and the death of all controls as a satisfactory indicator of both the safety and efficacy of a canine distemper vaccine batch.

While lethality may be the easier endpoint because of its objectivity and simplicity (Cussler et al. 1999), it is always worthwhile to identify reliable markers of predictive or impending mortality to serve as alternative and more humane endpoints. No purpose is served when the administration of a vaccine results in harm rather than protection but, as with all research studies and testing guidelines, there must be a balance between effective safety evaluation and humane endpoints for the sake of the laboratory animal.

HUMANE ENDPOINTS IN CANCER RESEARCH

Identification of humane endpoints in cancer research can be challenging. Although the wide range of tumor types and scientific objectives associated with this research prohibits standardization of humane endpoints (Wallace 1999, 2000), the United Kingdom Coordinating Committee on Cancer Research (UKCCCR) has developed a document to guide researchers working with animal models (UKCCCR 1988). Investigators should evaluate tumor size, tumor appearance, and animal condition to identify reliable indicators that may permit earlier termination of a study, and establish and validate endpoints that retain scientific objectives and avoid, minimize, or alleviate potential pain in the laboratory animals. Avoiding death or excessive tumor burden, particularly when coupled with clinical signs of pain or distress, should be a desirable goal in cancer research studies.

HUMANE ENDPOINTS IN PAIN RESEARCH

Of critical importance to this report, as well as to improvements in quality of life for both humans and animals, is research on pain itself, including the mechanisms of pain and methods of pain alleviation. Complicating the ethical issues inherent in producing pain in research subjects is the ability to accurately predict and measure pain responses in animals (Le Bars et al. 2001; Meyerson and Linderoth 2006; Walker et al. 1999). It is imperative for pain investigators to establish endpoints in each study design to minimize the duration and intensity of the pain and to validate those endpoints for the integrity, objectivity, and reproducibility of the study. Productive dialogue between the IACUC and researcher is critical for ensuring the best outcome for both the animals' welfare and the study objectives in these research programs (Mench 1999).

EUTHANASIA

Euthanasia, the act of inducing death without pain, is an acceptable method for relieving or alleviating pain that cannot be controlled by other means (NRC 1992, pp. 102-104). The humane death of an animal is one in which the animal is first rendered unconscious, and thus insensitive to pain, as rapidly as possible and with a minimum of fear and anxiety. A humane death, or endpoint, is a fundamental tenet of the US Principles for the Utilization and Care of Vertebrate Animals Used in Testing, Research, and Training (IRAC 2001), as Principle VI states that "[a]nimals that would otherwise suffer severe or chronic pain that cannot be relieved should be painlessly killed at the end of the procedure or, if appropriate, during the procedure."

There is no rigidly defined point at which euthanasia should be performed for humane reasons, as it is not possible to apply a single set of euthanasia criteria across all study designs, animal models, and experimental goals. The decision should involve a team approach among veterinarians, study directors, and animal care personnel using all available information about the affected animal(s). Body condition scores, as described in Chapter 3, can be used to determine when to consider euthanasia for humane reasons. The earliest possible indicators for euthanasia should be clearly identified so as to avoid pain and yet still achieve study objectives.

Methods of euthanasia have recently been updated by the American Veterinary Medical Association (AVMA 2007), although objective information on laboratory animals is sparse, particularly concerning the evaluation of potential pain and distress that may be caused by a particular euthanasia technique. The controversy that may result from this lack of data is evident in the recent discussions about the use of carbon dioxide on rodents (ACLAM 2005; AVMA 2007; Conlee et al. 2005; Hawkins et al. 2006; Kirkden et al. 2008; Niel et al. 2008; NRC 2003; Stephens et al. 2002). As conversations on this subject will likely continue, the reader is encouraged to follow the published literature for the most up-to-date information.

For all these reasons, well-designed objective studies of euthanasia across all laboratory animal species and age groups are needed and recommended. The assessment tools and measures to consider for such studies include electroencephalograms, electrocardiograms, electromyograms, arterial blood pressure, respiration and heart rates, serum biochemical parameters, pupil diameter, and behavioral changes. In particular, there is an urgent need for studies that provide measures of nociception, pain, distress, and the relation of these to loss of consciousness.

CONCLUSIONS AND RECOMMENDATIONS

Avoiding or minimizing pain in animal research is a fundamental obligation of all researchers for moral and ethical reasons. The criteria for early termination of a research project or alteration to a study design for the purpose of alleviating or avoiding pain in an animal are defined as humane endpoints. Identification and validation of humane endpoints should be considered for studies involving pain, but this is neither an easy nor a simple process.

1. It is important to ensure that endpoints are validated and based on sound science. Pilot studies are invaluable for the determination of earlier and more humane endpoints.
2. Given the wide scope of procedures and goals of animal research, no single reference can document every humane endpoint for every research protocol. Therefore, more effort must be made to identify appropriate humane endpoints for each. Good communication between researchers, veterinary staff, animal care staff, and the IACUC is crucial.
3. Productive strides have been made in the harmonization of safety assessment guidelines between countries but global harmonization is not yet complete. For global acceptance of humane endpoints in safety assessment test guidelines, dialogue should continue between all countries and agencies responsible for animal welfare, the environment, and human safety.
4. Efforts should continue in the development and validation of alternative procedures for incorporation in research projects and safety assessment tests to avoid or alleviate pain in laboratory animals.

Hendriksen and Morton (1999) observed that the goal of developing humane endpoints in animal experiments is constantly shifting. All scientists, managers, technicians, oversight committees, and regulators involved with animal experimentation where pain is a potential component should participate in regular communication and creative problem solving. The criteria for determining the humane end to a study should be frequently reevaluated and revised as new information becomes available. The sustained pursuit of these directed efforts can, and will, result in more humane animal use.

ADDENDUM

As stated in this chapter, the establishment of surrogate or humane endpoints as part of the experimental protocol and before experiments commence is one of the ways to minimize and alleviate pain and safeguard the well-being and welfare of laboratory animals. In support of this goal, two sample resources are provided for adaptation and use. The first is a score sheet to assess animals in cancer studies based on a behavioral and tumor scoring system (Table 5A-1). The recorded symptomatology will determine the diagnosis and measures for alleviation. The sheet can be adapted to any protocol or animal care facility system as long as the behavioral definitions are uniform across the same facility. The second resource is a model for developing guidelines for humane endpoints that may be suitable for any protocol within a facility (Box 5A-1).

TABLE 5A-1 Sample Tumor Scoring Sheet

				Calendar days								
SECTION A	**LESION/TUMOR CHARACTERISTICS**		**Score**							**MOUSE ID**	**Study No.**	**Group No.**
	Dry scab/crust forming coherent covering with skin		0									
	Acute burst releasing fluid/pus OR acute split at border		1									
	Chronically wet/weeping scab/crust OR solid yellow matter exposed		2									
	Bleeding or raw tissue exposed or white basal layer exposed		3									
	BEHAVIORAL CHANGES											
	No changes (i.e., normal)		0									
	Repeated grooming (tumor may not be easily visible or quite small)		1									
	Abnormal gait (tumor may not be easily visible or quite small)		1									
	Locomotion impeded (tumor is pronounced)		2									
	Recurrent scratching/biting of tumor		3									
	Nociception (struggling/sqeaking) on touching tumor		3									
IF YOU HAVE SCORED ANY 3s IN SECTION A, CULL TODAY.												
IF YOU HAVE SCORED ANY 2s IN SECTION A, REFER TO SECTION B, ELSE GO TO SECTION C.												
SECTION B	**SIZE/PROGRESSION OF LESION (expressed in orthogonal diameters**)**	**# D1/D2****										
	Shrinking		0									
	Static		1									
	2-3 mm growing		2									
	3-5 mm growing		3									
	5+ mm growing		4									
	SIZE/PROGRESSION OF TUMOR											
	Shrinking		0									
	Static		1									
	10-12 mm growing		2									
	12-14 mm growing		4									
	14+ mm growing (mean D1/D2 or D1 or either > 17 mm)		6									
	TOTAL FOR SECTION A + SECTION B											
	**# D1 and D2 are orthogonal diameters											
IF THE TOTAL FOR SECTIONS A + B IS 6, CULL WITHIN 1 DAY. IF 4-5 MONITOR DAILY, CULL WITHIN 1 WEEK IF NO IMPROVEMENT.												
ELSE GO TO SECTION C												
SECTION C	KEEP MONITORING AS REQUIRED (DAILY)											
		Date										

Adapted with permission from a scoring sheet developed by Fraser Darling, The Institute of Cancer Research, London, UK.

BOX 5A-1
Guidelines for Humane Endpoints in Animal Studies[a]

PURPOSE: To assure compliance with the Animal Welfare Act (AWA), the Guide for the Care and Use of Laboratory Animals (the "Guide") and (institutionally relevant) policies, as well as to promote good research. This policy describes the responsibilities and procedures that investigators and veterinary staff must follow when determining appropriate, humane endpoints.

PRINCIPLES: It is the responsibility of the Principal Investigator/Study Director (PI/SD) to define humane endpoints and to explore alternatives to death as an endpoint. If no alternative exists, the PI/SD should scientifically justify the use of death as an endpoint, and outline procedures that will be taken to minimize pain and distress to the animal.

Efforts should be made to minimize pain and distress experienced by animals used in research. This policy letter is to provide investigators with guidelines for determining humane endpoints in compliance with the XXXXXX policy. To this end, the use of death as an endpoint to experimental studies, rather than performing euthanasia to humanely terminate an animal, is discouraged and should be justified.

Each Animal Use Protocol (AUP), especially those that are anticipated to result in severe or chronic pain, should describe endpoint(s) and specify a plan and criteria for removal/euthanasia of animals from the study, or the disposition of animals at the termination of the study. For many studies, the endpoint will be euthanasia upon study completion, euthanasia at certain time points, or the return of animals to stock. For studies where moderate to severe clinical signs can be anticipated, the endpoint description in the AUP shall include identification of personnel responsible for decision making, specific criteria (body weight, mass size, appetite, etc.) that will be monitored at prescribed frequencies (daily, weekly, etc), and a disposition (treatment, euthanasia, early removal from study, etc.) once those criteria have been met or exceeded.

SCOPE: This policy covers any animal used for research.

POLICY STATEMENT: Animal studies may involve procedures that cause severe clinical signs or morbidity, and investigators should consider the selection of the most appropriate endpoint(s) for their study. This requires careful consideration of the scientific objectives of the study, the expected and possible adverse effects the research animals may experience, the most likely time course and progression of those adverse effects, and the earliest most predictive indicators of present or impending adverse effects. Prior to the initiation of the study, the PI/SD should determine the criteria that would lead to termination of the study for any animal, when appropriate, and the method of euthanasia to be employed. A clear chain of command for the decision-making process should be documented, including contingency plans if said individuals are unavailable for consultation. Optimally, studies are terminated when animals begin to exhibit severe clinical signs if this endpoint is compatible with meeting the research objectives. Such endpoints are preferable to death or moribundity (defined by the IACUC as imminent death) as endpoints since they minimize pain and distress.

There should be scientific justification in the AUP for allowing an animal to die without intervention if the goals of a study can be accomplished by euthanizing animals before they become moribund.

Animals involved in experiments that may lead to moribundity or death should be monitored daily (including weekends) by personnel experienced in recognizing signs of morbidity. Once severe clinical signs develop, more frequent observation (2-3 times daily) may be required.

The following conditions usually necessitate euthanasia. The PI/SD should provide scientific justification for exemptions:

- Rapid weight loss of ≥20% of body weight.
- Extended period of weight loss, progressing to emaciated state.
- Surgical complications unresponsive to medical intervention.
- Combination of the following: poor physical appearance (very rough hair coat, abnormal posture, grunting on exhalation); abnormal behavior (reduced mobility/unconsciousness, unsolicited vocalizations, self-mutilation); severe depression or abnormal/exaggerated responses to external stimuli.
- Severe respiratory distress, which is unresponsive to treatment.
- Occurrence of a serious injury or trauma from which recovery is unlikely.
- Neurological signs (e.g., persistent convulsions, persistent circling, paresis/paralysis) that interfere with eating and drinking and from which recovery is unlikely.
- Frank bleeding from any orifice, which is unresponsive to treatment.
- One or more skin ulcers that do not heal, depending upon the species and severity of the ulcers.
- Mass size or location that interferes with normal function or ulcerates with no evidence of healing.
- A mass that is greater than 15% of normal body weight. For chronic toxicology studies (e.g., 2-year carcinogenicity studies), it is necessary to rely on experience and good judgment when deciding when to euthanize an animal as a result of one or more masses. Many of these masses grow slowly and do not compromise the animal.

RESPONSIBILITY: The PI/SD is responsible for ensuring that this IACUC policy is followed. Exceptions to this policy should be scientifically justified and approved by the IACUC before they can be implemented.

The IACUC has the authority, mandated by law (7 U.S. Code Section 2131 et seq.), to act on behalf of the head of the institution to investigate and if necessary suspend any activity which violates applicable laws, regulations, standards, guidelines, policies and procedures.

REFERENCES

Montgomery CA. 1990. Oncologic and toxicologic research: Alleviation and control of pain and distress in laboratory animals. Canc Bull 42(4):230-237.

Stokes WS. 2002. Humane endpoints for laboratory animals used in regulatory testing. ILAR J 43(Suppl):S31-S38.

---

[a]Example of an IACUC Humane Endpoints Policy Letter, developed by Maryfrances Lutz, Cecilia Pate, and Gaye Ruble, sanofi-aventis US.

REFERENCES

ACLAM (American College of Laboratory Animal Medicine). 2005. Public Statements: Report of the ACLAM Task Force on Rodent Euthanasia. Available at www.aclam.org/print/report_rodent_euth.pdf. Accessed January 5, 2008.

AVMA (American Veterinary Medical Association). 2007. AVMA Guidelines on Euthanasia. Available at www.avma.org/issues/animal_welfare/euthanasia.pdf

AWIC (Animal Welfare Information Center). Humane Endpoints. Available at http://awic.nal.usda.gov/nal. Accessed January 5, 2008.

Carmichael NG, Barton HA, Boobis AR, Cooper RL, Dellarco VL, Doerrer NG, Fenner-Crisp PA, Doe JE, Lamb JC 4th, Pastoor TP. 2006. Agricultural chemical safety assessment: A multisector approach to the modernization of human safety requirements. Crit Rev Toxicol 36(1):1-7.

Castle P. 1999. The European pharmacopoeia and humane endpoints. In: Hendriksen CFM, Morton DB, eds. Humane Endpoints in Animal Experiments for Biomedical Research, Proceedings of the International Conference, 22-25 November 1998, Ziest, The Netherlands. The Royal Society Medical Press. pp. 15-19.

CCAC (Canadian Council on Animal Care). 1998. Guidelines on choosing an appropriate endpoint in experiments using animals for research, teaching and testing. Available at www.ccac.ca/en/CCAC_Programs/Guidelines_Policies/GDLINES/ENDPTS/g_endpoints.pdf. Accessed January 5. 2008.

CFR (Code of Federal Regulations). 2006. 9 CFR 113.33. Available at www.access.gpo.gov/nara/cfr/waisidx_00/9cfrv1_00.html. Accessed January 5 2008.

CFR. 2008. Regulations 21 CFR 610.11. Available at www.accessdata.fda.gov/scripts/cdrh/cfdocs/cfcfr/CFRSearch.cfm?fr=610.11. Accessed October, 26 2008.

Collins FS, Gray GM, Bucher JR. 2008. Transforming environmental health protection. Science 319:906-907.

Conlee KM, Stephens ML, Rowan AN, King LA. 2005. Carbon dioxide for euthanasia: Concerns regarding pain and distress, with special reference to mice and rats. Lab Anim 39:137-161.

Cooper RL, Lamb JC, Barlow SM, Bentley K, Brady AM, Doerrer NG, Eisenbrandt DL, Fenner-Crisp PA, Hines RN, Irvine LF, Kimmel CA, Koeter H, Li AA, Makris, SL, Sheets LP, Speijers G, Whitby KE. 2006. A tiered approach to life stages testing for agricultural chemical safety assessment. Crit Rev Toxicol 36(1):69-98.

Copps J. 2005. Issues related to the use of animals in biocontainment research facilities. ILAR J 46(1):34-43.

Cussler K, Morton DB, Hendriksen CFM. 1999. Humane endpoints in vaccine research and quality control. In: Hendriksen CFM, Morton DB, eds. Humane Endpoints in Animal Experiments for Biomedical Research, Proceedings of the International Conference, 22-25 November 1998, Ziest, The Netherlands. The Royal Society Medical Press. pp. 95-101.

DeHaven WR. 2002. Best practices for animal care committees and animal oversight. ILAR J 43(Suppl):S59-S62.

Doe JE, Boobis AR, Blacker A, Dellarco V, Doerrer NG, Franklin C, Goodman JI, Kronenberg JM, Lewis R, McConnell EE, Mercier T, Moretto A, Nolan C, Padilla S, Phang W, Solecki R, Tillbury L, van Ravenzwaay B, Wolf DC. 2006. A tiered approach to systemic toxicity testing for agricultural chemical safety assessment. Crit Rev Toxicol 36(1):37-68.

Durham RA, Sawyer DC, Keller WF, Wheeler CA. 1992. Topical ocular anesthetics in ocular irritancy testing: A review. Lab Anim Sci 42(6):535-541.

ECHA (European Chemicals Agency). 2008. REACH Guidance. Available at http://guidance.echa.europa.eu/. Accessed January 2, 2009.

Federal Register Notice, Volume 72, Number 89, May 9, 2007. Department of Health and Human Services, National Toxicology Program (NTP), NTP Interagency Center for the Evaluation of Alternative Toxicological Methods (NICEATM): Request for Data on the Use of Topical Anesthetics and Systemic Analgesics for In Vivo Eye Irritation Testing. Available at http://iccvam.niehs.nih.gov/SuppDocs/FedDocs/FR/FR_E7_8898.pdf.

Gallo MA. 2001. History and scope of toxicology. In: Klaassen CD, ed. Casarett and Doull's Toxicology: The Basic Science of Poisons, 6th Edition, New York, McGraw Hill Medical Publishing Division, pp. 3-10.

Hawkins P, Playle L, Golledge H, Leach M, Banzett R, Coenen A, Cooper J, Danneman P, Flecknell P, Kirkden R, Niel L, Raj M. 2006. Newcastle Consensus Meeting on Carbon Dioxide Euthanasia of Laboratory Animals. Available at www.nc3rs.org.uk/downloaddoc.asp?id=416&page=292&skin=0.

Hendriksen CFM. 2002. Refinement, reduction, and replacement of animal use for regulatory testing: Current best scientific practices for the evaluation of safety and potency of biologicals. ILAR J 43(Suppl):S43-S48.

Hendriksen CFM, Morton DB, eds. 1999. Humane Endpoints in Animal Experiments for Biomedical Research. Proceedings of the International Conference, 22-25 November 1998, Ziest, The Netherlands. London: The Royal Society Medical Press.

Hendriksen CFM, Steen B. 2000. Refinement of vaccine potency testing with the use of humane endpoints. ILAR J 41:105-113.

Hicks JM. 1997. Animal welfare and toxicology/safety studies: Making sense of the regulatory environment. Cont Topics 36(3):49-54.

ILAR (Institute for Laboratory Animal Research). 2000. Humane Endpoints for Animals Used in Biomedical Research and Testing. ILAR J 41(2). Available at http://dels.nas.edu/ilar_n/ilarjournal/41_2/.

ILAR. 2009. Mission and Core Values. Available at http://del.nas.edu/ilar_n/ilarhome/mission.shtml.

ILSI-HESI (International Life Sciences Institute–Health and Environmental Sciences Institute). 2008. Agricultural Chemical Safety Assessment factsheet. Available at http://www.hesiglobal.org/files/public/Factsheets/AgriculturalChemicalSafetyAssessment.pdf. Accessed July 14, 2009.

IRAC (Interagency Research Animal Committee). 1985. The U.S. Government Principles for the Utilization and Care of Vertebrate Animals Used in Testing, Research, and Training. Federal Register Vol. 50, No. 97 (May 20, 1985). Available at http://grants.nih.gov/grants/olaw/references/phspol.htm#USGovPrinciples. Accessed June 9, 2008.

Jaax J. 2005. Administrative issues related to infectious disease research in the age of bioterrorism. ILAR J 46(1):34-43.

JMAFF (Japanese Ministry of Agriculture, Forestry and Fisheries). 2000. Testing Guidelines for Toxicology Studies.

Kirkden RD, Niel L, Stewart SA, Weary DM. 2008. Gas killing of rats: The effect of supplemental oxygen on aversion to carbon dioxide. Anim Welf 17:79-87.

Le Bars D, Gozariu M, Cadden SW. 2001. Animal models of nociception. Pharmacol Rev 53:597-652.

Mench J. 1999. Defining endpoints: The role of the animal care committee. In: Humane Endpoints in Animal Experiments for Biomedical Research. London: The Royal Society Medical Press. pp. 133-138.

Merrill RA. 2001. Regulatory toxicology. In: Klaassen CD, ed. Casarett and Doull's Toxicology: The Basic Science of Poisons, 6th ed. New York: McGraw Hill Medical Publishing Division. pp. 1141-1153.

Meyerson BA, Linderoth B. 2006. Mode of action of spinal cord stimulation in neuropathic pain. J Pain Sympt Manag 31(4S):S6-S12.

Morton DB. 1999. Humane endpoints in animal experimentation for biomedical research: Ethical, legal and practical aspects. In: Humane Endpoints in Animal Experiments for Biomedical Research. London: The Royal Society Medical Press. pp. 5-12.

Morton DB. 2000. A systematic approach for establishing humane endpoints. ILAR J 41(2): 80-86.

Morton DB, Burghardt GM, Smith JA. 1990. Critical anthropomorphism, animal suffering and the ecological context. Hastings Cent Rep 20(3):13-19.

Nemzek JA, Xiao H-Y, Minard AE, Bolgos GL, Remick DG. 2004. Humane endpoints in shock research. Shock 21(1):17-25.

Nemzek JA, Hugunin KM, Opp MR. 2008. Modeling sepsis in the laboratory: Merging sound science with animal well-being. Comp Med 58(2):120-128.

Niel L, Stewart SA, Weary DM. 2008. Effect of flow rate on aversion to gradual-fill carbon dioxide exposure in rats. Appl Animl Behav Sci 109:77-84.

Nies AS. 2001. Principles of Therapeutics. In: Hardman JG, Limbird LE, eds. Goodman and Gilman's The Pharmacological Basis of Therapeutics, 10th ed. New York: McGraw-Hill. pp. 45-66.

NIH (National Institutes of Health). 1997. Validation and Regulatory Acceptance of Toxicological Test Methods: A Report of the ad hoc Interagency Coordinating Committee on the Validation of Alternative Methods. NIH Publication No. 97-3981, Research Triangle Park, NC.

NIH. 2008. NIH Collaborates with EPA to Improve the Safety Testing of Chemicals. News Press Release. Available at www.nih.gov/news/health/feb2008/nhgri-14.htm.

NIH/USEPA (National Institutes of Health/US Environmental Protection Agency). 2008. MOU (Memorandum of Understanding) on High Throughput Screening, Toxicity Pathway Profiling, and Biological Interpretation of Findings. Available at www.niehs.nih.gov/news/releases/2008/docs/ntpncgcepamou.pdf.

NRC (National Research Council). 1992. Recognition and Alleviation of Pain and Distress in Laboratory Animals. Washington: National Academy Press.

NRC. 2003. Guidelines for the Care and Use of Mammals in Neuroscience and Behavioral Research. Washington: National Academies Press.

NRC. 2004. The Development of Science-based Guidelines for Laboratory Animal Care: Proceedings of the November 2003 International Workshop. Washington: National Academies Press.

NRC. 2007. Toxicity Testing in the Twenty-first Century: A Vision and a Strategy. Washington: National Academies Press.

NRC. 2008. Recognition and Alleviation of Distress in Laboratory Animals. Washington: National Academies Press.

OECD (Organization for Economic Cooperation and Development). 1987. Guidelines for Testing of Chemicals, Section 4: Health Effects. Paris: OECD.

OECD. 2000. Guidance Document on the Recognition, Assessment, and Use of Clinical Signs as Humane Endpoints for Experimental Animals Used in Safety Evaluation. Paris: OECD.

OLAW/ARENA (Office of Laboratory Animal Welfare/Applied Research Ethics National Association). 2002. Institutional Animal Care and Use Committee Guidebook, 2nd ed. Bethesda: National Institutes of Health.

Olfert ED, Godson DL. 2000. Humane endpoints for infectious disease animal models. ILAR J 41(2):99-104.

Patrone G, Sacca SC, Macri A, Rolando M. 1999. Evaluation of the analgesic effect of 0.1% indomethacin solution on corneal abrasions. Ophthalmologica 213:350-354.

Peyman GA, Rahimy MH, Fernandes ML. 1994. Effects of morphine on corneal sensitivity and epithelial wound healing: Implications for topical ophthalmic analgesia. Br J Ophthalmol 78:138-141.

Sass N. 2000. Humane endpoints and acute toxicity testing. ILAR J 41(2). Available at http://dels.nas.edu/ilar_n/ilarjournal/41_2/AcuteToxicity.shtml.

Stephens ML, Conlee K, Alvino G, Rowan AN. 2002. Possibilities for refinement and reduction: Future improvements within regulatory testing. ILAR J 43(Suppl):S74-S79.

Stiles J, Honda CN, Krohne SG, Kazacos EA. 2003. Effect of topical administration of 1% morphine sulfate solution on signs of pain and corneal wound healing in dogs. Am J Vet Res 64 (7):813-818.

Stokes WS. 2002. Humane endpoints for laboratory animals used in regulatory testing. ILAR J 43(Suppl):S31-S38.

Stokes WS. 2005. Summary of ICCVAM-NICEATM-ECVAM Ocular Toxicity Scientific Symposia. Presentation. Available at http://iccvam.niehs.nih.gov/meetings/SACpresent/SACocular.pdf. Accessed July 14, 2009.

UKCCCR (United Kingdom Coordinating Committee on Cancer Research). 1988. Guidelines for the welfare of animals in experimental neoplasia. Lab Anim 22:195-201.

USEPA (United States Environmental Protection Agency). 2008. Summary of the Toxic Substances Control Act, 15 USC 2601 et seq. (1976). Available at www.epa.gov/lawsregs/laws/tsca.html. Accessed January 5 2008.

USEPA OPPTS (Office of Prevention, Pesticides, and Toxic Substances). 1998. 870 Series Final Guidelines: Health Effect Test Guidelines. Available at www.epa.gov/opptsfrs/publications/OPPTS_Harmonized/870_Health_Effects_Test_Guidelines/Series/. Accessed January 5, 2008.

USEPA OPPTS. 2002. 870 Series Final Guidelines: Health Effect Test Guidelines, OPPTS 870.1100. Acute Oral Toxicity. Available at www.epa.gov/opptsfrs/publications/OPPTS_Harmonized/870_Health_Effects_Test_Guidelines/Series/. Accessed July 14, 2009.

Walker K, Fox AJ, Urban LA. 1999. Animal models of pain. Mol Med Today 5:319-321.

Wallace J. 1999. Humane endpoints in cancer research. In: Hendriksen CFM, Morton DB, eds. Humane Endpoints in Animal Experiments for Biomedical Research. Proceedings of the International Conference, 22-25 November 1998. Ziest: The Netherlands. The Royal Society Medical Press. pp. 79-84.

Wallace J. 2000. Humane endpoints in cancer research. ILAR J 41(2):87-93.

Appendixes

APPENDIX

A

Models of Pain

INTRODUCTION

Pain can be characterized by its duration (from momentary to chronic), location (e.g., muscle, viscera), or cause (e.g., nerve injury, inflammation). Characterization of pain by duration may be arbitrary (i.e., when does pain become chronic?), but is useful because most significant human pain conditions are long-lasting, whether referred to as persistent or chronic.

Numerous animal models exist for the exploration of mechanism(s) and mediators of persistent pain in particular. The principal rationale for developing and using such models is that the sources and mechanisms of momentary pain differ significantly from those of persistent pain. Knowledge of these mechanisms is necessary to address the second objective of such studies, namely the development of (usually) pharmacological strategies for targeted, improved pain management.

Table A-1 presents commonly used models of persistent pain in animals and the subsequent sections provide an overview of response measures and other features of these models. Most of the models were developed in rodents (rats or mice), unless otherwise specified, and behavioral and other response measures are described for these species alone. Momentary, stimulus-evoked pain is not discussed because stimulus duration is typically short, responses are generally reflexive in nature (e.g., tail withdrawal), and the stimulus intensity is not injurious to tissue. Animal models of momentary pain are fully described in a comprehensive review by LeBars and colleagues (2001).

TABLE A-1 Animal Models of Persistent Pain[a]

Type of Pain Model	Insult	References
Inflammatory pain models		
Hindpaw		Hong and Abbott 1994
	carrageenan	Honoré et al. 1995
	zymosan	Meller and Gebhart 1997
	complete Freund's adjuvant (CFA)	Iadarola et al. 1988
	bee venom	Lariviere and Melzack 1996
	formalin	Dubuisson and Dennis 1997 Hunskaar and Hole 1987 Allen and Yaksh 2004
	capsaicin	Caterina et al. 2000
	ultraviolet-B irradiation	Bishop et al. 2007
Joints		
	cruciate ligament transection	Vilensky et al. 1994
	intra-articular (arthritis)	Sluka and Westlund 1993 Bendele et al. 1999 Neugebauer et al. 2007
	collagen-induced arthritis	Brand et al. 2004
Neuropathic pain models		
Central nervous system		
	spinal cord trauma (blunt)	Young 2002
	spinal cord insult (chemical)	Yezierski et al. 1998
	experimental allergic encephalomyelitis	Olechowski et al. 2009
Peripheral nervous system		
	mononeuropathies (chronic constriction injury)	Bennett and Xie 1988
	spinal nerve ligation/transection	Kim and Chung 1992
	spared nerve preparation	Decosterd and Woolf 2000 Shields et al. 2003
	partial nerve ligation/transection	Seltzer et al. 1990 Malmberg and Basbaum 1998 Aley et al. 1996 Polomano et al. 2001 Smith et al. 2004
	dorsal root ganglion compression	Hu and Xing 1998
	complex regional pain syndrome (CRPS)	Coderre et al. 2004
	streptozotocin-induced diabetic neuropathy	Rakieten et al. 1963 Wuarin-Bierman et al. 1987
	HIV (gp120)/antiretrovirals	Wallace et al. 2007
	herpes zoster/postherpetic neuralgia	Sadzot-Delvaux et al. 1990
Visceral pain models[b]		
	stomach (ulceration, gastritis)	Ozaki et al. 2002 Lamb et al. 2003

Type of Pain Model	Insult	References
	urinary bladder (cyclophosphamide, zymosan)	Lantéri-Minet et al. 1995 Randich et al. 2006a,b
	colon (acetic acid, trinitrobenzesulfonic acid, zymosan)	Morris et al. 1989 Burton and Gebhart 1995 Coutinho et al. 1996 Al-Chaer et al. 2000 Kamp et al. 2003 Jones et al. 2007
	ureteral calculosis	Giamberardino et al. 1995
	pancreatitis	Vera-Portocarrero et al. 2003
	female reproductive organs	Wesselmann et al. 1998 Berkley et al. 1995, 2007
Muscle pain models		
	intramuscular injection (chemical)	Radhakrishnan et al. 2003 Sluka et al. 2001
Postoperative (incisional) pain models		
	glabrous skin	Brennan et al. 1996 Banik et al. 2006
	hairy skin	Duarte et al. 2005
Orofacial pain models		
	inferior alveolar nerve or infraorbital nerve ligation	Vos et al. 1994 Tsuboi et al. 2004
	tooth preparation	Law et al. 1999
	orofacial inflammation	Clavelou et al. 1995 Morgan and Gebhart 2008
	temporomandibular joint inflammation	Hartwig et al. 2003
Models of head pain (headache, migraine)		
	subarachnoid blood	Ebersberger et al. 1999
	chemical irritation of the dura (inflammatory soup)	Burstein et al. 1998
	traumatic head injury	Browne et al. 2006
Burn models		
	skin (52°C thermal stimulation for 45 sec to anesthetized rat)	Nozaki-Taguchi and Yaksh 1998 Allen and Yaksh 2004
Cancer pain models[c]		
	bone cancer	Schwei et al. 1999
	pancreatic cancer	Lindsay et al. 2005
	review of animal models	Pacharinsak and Beitz 2008

[a]Most of these models are provided here for completeness and are not discussed further in this report.

[b]Many of these models are inflammatory in nature, but response measures differ significantly from nonvisceral inflammatory models.

[c]These models are likely associated with both inflammation and nerve injury.

ANIMAL MODELS OF PERSISTENT PAIN

Inflammatory Pain Models

Rodent hindpaw inflammation is a commonly used model of persistent inflammatory pain in which noxious stimuli are applied to the glabrous (thermal) or glabrous and hairy (mechanical) skin of the hindpaw. Response measures are typically hindpaw withdrawal latency to heat (seconds) or mechanical withdrawal threshold (g or mN). Once baseline response measures have been determined, an inflammogen is injected into either the dorsal hairy or ventral glabrous skin and withdrawal responses are assessed over time (hours to days). Post-treatment response measures are hyperalgesic, meaning that response latency to heat is faster and mechanical withdrawal thresholds (typically assessed using von Frey-like nylon monofilaments, each of which has a different bending force) are lower. Edema, which is also a consequence of such an injection, is greatest after the injection of carrageenan (or carrageenan plus kaolin) and least following complete Freund's adjuvant (CFA). The nature and duration of hyperalgesia differ between the inflammogens—some produce greater thermal hyperalgesia and others greater mechanical hyperalgesia. The hyperalgesia produced by carrageenan is typically assessed over 4 to 6 hours but can persist more than 24 hours, whereas that produced by CFA peaks at 1 to 2 days, although it may remain present for more than 1 week, during which it decreases.

Hindpaw injection of formalin or capsaicin is also used to assess intense, short-lasting (minutes to tens of minutes) persistent pain. The effect of formalin is concentration-dependent (Kaneko et al. 2000; Saddi and Abbott 2000) and is expressed by hindlimb licking and shaking that occur principally in two phases. The first phase is short (~10 min), followed by a brief (~5 min) period of relative quiescence, after which a second phase of hindlimb shaking and licking lasts an additional 50 minutes or so. The formalin test has also been characterized in infant rats (Abbott and Guy 1995). Capsaicin selectively activates a subset of nociceptors that express the transient receptor potential vanilloid receptor (TRPV1), an ion channel that responds to capsaicin, protons, and heat. Intradermal injection of capsaicin produces a relatively short-lasting (minutes) but intense pain associated with hyperalgesia that persists for hours after the capsaicin-produced pain has resolved.

Joint Inflammation Models

There are physical, chemical, and biologic methods to produce inflammatory states that mimic painful conditions of joints. Among physical methods, anterior (cranial) cruciate ligament transection produces instability of the knee joint and is a common model of osteoarthritis in dogs and rabbits. Immediately after ligament disruption, animals exhibit joint swelling as well as a dramatic reduction in weight bearing on the unstable limb although there will be a return to some degree of weight bearing accompanied by chronic joint instability.

Chemical methods include the intra-articular injection of inflammogens (e.g., kaolin, carrageenan, iodoacetate, collagenase, urate crystals) to cause synovitis, varying degrees of cartilage destruction and subsequent joint swelling, lameness, and decreased activity. Hyperalgesia develops rapidly (within 4 hours); both inflammation and the duration of inflammation depend on the agent and dose.

An example of a biologic model is antigen-induced arthritis, which develops after intra-articular injection of a protein antigen against which animals have been previously immunized (e.g., methylated bovine serum albumin). The condition appears only in the injected joints, as soon as 3 to 5 days after injection. The acute form of this arthritis is characterized by joint and soft-tissue swelling, reduced weight bearing, and altered activity until the joint swelling declines, typically after 1 week. A longer-lasting chronic arthritis model (30 to 300 days), established after intra-articular antigen, involves reactivation of arthritis (arthritis flare) by reinjection 1 month later (Moran and Bogoch 1999; van den Berg et al. 2007).

Models of rheumatoid arthritis entail activation of an immune response that targets multiple joints. One example is adjuvant arthritis, a polyarticular disease that develops 10 to 45 days after intravenous or intraperitoneal injection of CFA and typically resolves over a month. Another example is collagen-induced arthritis produced by immunizing animals with type II collagen; the time course of the resulting arthritis differs between rats and mice, but onset generally occurs 2 to 4 weeks after immunization. Resolution of clinical signs occurs in rats after 30 to 45 days, whereas susceptible mice demonstrate disease 8 to 12 weeks postimmunization. The duration, severity, and location of arthritis after collagen immunization depends on the genetic background of the animals being used as well as the source of the collagen (autologous vs. heterologous) (Griffiths et al. 2007; van den Berg et al. 2007).

In general, pain associated with inflammatory joint models is assessed by documenting changes in body weight, joint circumference, joint mobility, degree of weight bearing, soft tissue swelling, general activity, and gait. In addition, investigators often quantify latency to withdrawal or vocalization in response to pressure applied across the joint or, as a model of secondary hyperalgesia, responses to heat or mechanical stimulation of the hindpaw.

Visceral Pain Models

Although once considered models of visceral pain, irritants such as acetic acid, hypertonic saline, phenylquinone, and others injected intraperitoneally do not selectively act on the viscera, and moreover produce a behavior (writhing) that is inescapable. Accordingly, such models have fallen into disfavor and have been largely replaced with hollow organ balloon distension, which reproduces in humans the quality, location, and intensity of actual visceral pain (Ness and Gebhart 1990). Methods for distension of rat stomach (Ozaki et al. 2002), rat (Ness et al. 2001) and mouse (Ness and Elhefni 2004) urinary bladder, and rat (Gebhart and Sengupta 1996) and mouse (Christianson and Gebhart 2007) colon have been fully described.

Hollow organ distension produces several quantifiable responses, including contraction of skeletal (nonvisceral) muscles (termed the visceromotor response) and increases in blood pressure and heart rate. Electromyographic (EMG) recordings of muscle contraction, which require the surgical implantation of EMG recording electrodes in appropriate muscles, generally provide the most reliable response measure. Blood pressure and heart rate measurement require either surgical implantation of an arterial catheter, which can be difficult to keep patent in rodents, or expensive telemetric methods for long-term recording of these measures. These responses to organ distension are organized in the brainstem (and thus are not simple nociceptive reflexes) and are best assessed in unanesthetized animals because anesthetic drugs affect responses (e.g., pressor effects are converted to depressor effects; Ness and Gebhart 1990).

Because nonulcer dyspepsia, interstitial cystitis/painful bladder syndrome, and inflammatory and irritable bowel syndromes are relatively common human diseases for which management of pain is poor, many models entail the irritation or inflammation of hollow organs to assess the mechanisms underlying the hypersensitivity that characterizes these human disorders.[1] The following models have been developed to study these mechanisms:

- lower esophageal irritation (usually with HCl), stomach ulceration (acetic acid-produced lesions), and inflammation (oral ingestion of 0.1% iodoacetic acid; Ozaki et al. 2002),
- colon inflammation (e.g., intracolonic trinitrobenzenesulfonic acid or acetic acid), hypersensitivity in the absence of inflammation (intracolonic zymosan; Jones et al. 2007),
- urinary bladder inflammation (intraperitoneal administration of

[1] As indicated in Chapter 2 (see Ontogeny of Pain), organ insult or stress (e.g., maternal separation) in early life can lead to visceral hypersensitivity in adults (Al-Chaer et al. 2000; Coutinho et al. 2002; Randich et al. 2006a).

cyclophosphamide, which is metabolized to the bladder irritant acrolein and produces cystitis; Lanteri-Minet et al. 1995), and
- uterine inflammation (Wesselmann et al. 1998).

In unanesthetized rodents, baseline responses to balloon distension are acquired before organ insult and monitored over time (days to weeks) after the insult, when they are typically exaggerated (increased) and occur at reduced response thresholds (i.e., they are hyperalgesic or hypersensitive).

Inflammatory models of the pancreas have also been developed (e.g., Vera-Portocarrero et al. 2003). The response measure in these models is typically mechanical hypersensitivity (e.g., von Frey probing) determined in the area of referred sensation (thorax and abdominal skin). Similarly, one response measure in a kidney stone (ureteral calculosis) model is mechanical hypersensitivity, including of the paraspinous muscles. This model is also associated with episodes of lordosis-like stretching and hunching, which can be quantified by frequency as well as intensity (Giamberardino et al. 1995).

Postoperative (Incisional) Pain Models

Models of postoperative pain have revealed that the mechanisms and subsequent control of postoperative pain differ significantly from those of inflammatory pain. These models involve an incision of glabrous or hairy skin of controlled length and depth to determine the relative contributions of skin, fascia, and underlying muscle to postoperative pain. To eliminate any possible contribution of infection, the incisions are made under aseptic conditions. Response measures include both thermal (heat) and mechanical (von Frey probing) hyperalgesia at (primary hyperalgesia) and adjacent to (secondary hyperalgesia) the incision. An incision of glabrous hindpaw skin and fascia leads to both thermal and mechanical hyperalgesia that is maximal within the first 24 to 48 hours after incision and typically lasts 3 to 4 days. When underlying muscle is included in the incision, the duration (but not the magnitude) of hyperalgesia is usually extended by 1 day.

Orofacial Pain Models

The injection of inflammogens into the temporomandibular joint (TMJ) or subcutaneous tissues of the face produces models of orofacial pain. Injection of mustard oil into the TMJ causes rapid onset of swelling and behavioral changes—initially, freezing behavior, followed by a second phase of active behaviors such as facial rubbing or grooming, chewing movements, and head shaking. These active behaviors peak at 1.5 to 2 hours and return to baseline by 5 hours after the injection (Hartwig et al. 2003). Subcutaneous formalin injection into the facial whisker pad results in acute onset of facial rubbing in rats that lasts at least 45 minutes. The duration of grooming activity and edema after formalin injection is concentration dependent (Clavelou et al. 1995). Whisker pad injection of CFA produces a longer-lasting (2 weeks) thermal and mechanical orofacial hyperalgesia (Morgan and Gebhart 2008).

Transection or injury of the trigeminal nerve is commonly used to model neuropathic pain of the face and mouth. Transection of the inferior alveolar nerve, a branch of the trigeminal nerve, produces mechanical allodynia in rats after 2 to 3 days (Tsuboi et al. 2004). Similarly, nerve constriction results in nerve injury and mechanical hyperalgesia. Unilateral chronic constriction injury (CCI) has been used in rats to study orofacial allodynia. After unilateral loose ligation of the infraorbital nerve, rats develop a biphasic behavioral response. In the early postligature phase (days 1 to 15), they demonstrate increased grooming activity at the site of nerve injury but are *hypo*responsive to mechanical stimuli; on postconstriction days 15 to 130, the rats become *hyper*responsive to mechanical stimuli, demonstrating maximal escape responses to all stimulus intensities. Decreased weight gain and altered activity also occur in this constriction injury model (Vos et al. 1994).

Muscle Pain Models

Models of persistent muscle pain include intramuscular injection of carrageenan or acidic saline. Unilateral injection of carrageenan into the gastrocnemius muscle of rats produces acute inflammation with edema and reduced withdrawal latencies in the first 4 to 24 hours. Hyperalgesia also develops in the contralateral limb 1 to 2 weeks after injection, suggesting involvement of central nervous system mechanisms. Mechanical and thermal hyperalgesia are dependent on the concentration of carrageenan and may last 7 to 8 weeks (Radhakrishnan et al. 2003).

Injection of acidic saline in the gastrocnemius produces secondary mechanical but not thermal hyperalgesia (in tests on the hindpaw). The magnitude and contralateral spread of hyperalgesia are directly related to acidity and also depend on the timing of repeated intramuscular injections. Despite the reductions in mechanical threshold caused by acidic saline injection, changes do not appear in either behavior (i.e., gait and weight bearing remain normal, and there is no limb guarding) or muscle histology (Sluka et al. 2001).

Neuropathic Pain Models

Of the two major classes of clinical pain conditions—those produced by tissue injury and those produced by nerve injury—the latter for many years were very difficult to model in animals. The human clinical condition can result from traumatic, metabolic, or drug-induced injury to either the peripheral nervous system (e.g., diabetic neuropathy, postherpetic neuralgia, complex regional pain syndrome, or chemotherapy-induced neuropathy) or the CNS (e.g., from multiple sclerosis, stroke-induced destruction of tissue, or spinal cord injury). Although there have been many attempts (e.g., the use of streptozotocin to produce an animal model of diabetes and its associated neuropathy) to model the different clinical conditions, most studies have built on the principle that neuropathic pain arises from partial nerve injury (e.g., of a peripheral nerve) or abnormal neuronal activity.

The first model of pain induced by nerve injury (Bennett and Xie 1988) demonstrated that constriction of the sciatic nerve of the rat leads to persistent pain with significant mechanical and thermal (warm and cold) hypersensitivity as well as signs of recurrent spontaneous pain. Researchers inferred the latter from the animals' apparent protection of the partially denervated hindlimb. There have been many variations of this model, and they are commonly used largely because they are highly reproducible and involve a relatively short surgical procedure. Among these are models in which (1) one-half to two-thirds the diameter of the sciatic nerve is cut (Seltzer et al. 1990), (2) one or two spinal nerves (usually L5 and L6) are ligated and/or cut just distal to the dorsal root ganglion (Kim and Chung 1992), and (3) two of the three branches of the sciatic nerve are cut distal to its trifurcation (Decosterd and Woolf 2000). In general, these models are associated with a more pronounced mechanical allodynia than heat hyperalgesia; cold hypersensitivity is prominent. These models were developed in the rat and, importantly, several have been adapted for the mouse, which has proven very valuable for the study of the genetic basis of different nerve injury-induced pain conditions (Malmberg and Basbaum 1998; Shields et al. 2003).

Although spontaneous pain may be associated with these models (see below), this is not readily apparent and is certainly difficult to document. There is rarely any significant change in behavior or weight loss that might indicate ongoing pain. Thus testing of the animals typically involves assessment of changes in mechanical paw withdrawal thresholds (using von Frey-like nylon monofilaments or the Randall Selitto apparatus) and paw withdrawal latencies for assessment of heat hyperalgesia. Cold hypersensitivity is very difficult to assess in rodents. Some laboratories rely on the evaporation of acetone applied to the affected hindpaw; the endpoint is shaking of the paw. Responses on a single cold plate are often used, but typically very cold temperatures are necessary in order to generate any behavioral response. For this reason, better results are reported using a two-plate method in which an animal can escape to the plate that is less cold.

The reliability of these different approaches to modeling neuropathic pain is evident primarily from the demonstration that drugs that are effective (or not) in the clinic for neuropathic pain are effective in the animal models. For example, many anticonvulsant drugs, which either block sodium channels or enhance GABAergic inhibitory tone, are effective in the animal models and also are the mainstay for neuropathic pain relief in humans. In contrast, there is general agreement that nonsteroidal anti-inflammatory drugs are quite ineffective in humans with neuropathic pain, and the same is true in the animal models. Opioids also are less effective in neuropathic pain models than in inflammatory models, and this is commonly observed in the clinic.

As noted above, one of the problematic adverse side effects of chemotherapy treatment for cancer pain is the development of a profound peripheral neuropathy with mechanical allodynia, thermal hypersensitivity, and ongoing, often burning pain. In recent years several laboratories have developed neuropathic pain models based on treatment with vincristine or taxol; the treatment typically involves weeks of drug administration to gradually produce in the animals a significant mechanical and thermal hypersensitivity to both warm and cold stimuli (the hypersensitivity disappears when the drug treatment ends). Very recently, a somewhat comparable condition has been reported following the administration of antiretroviral drugs, which are used in the treatment of HIV and are also often associated with the development of severe neuropathic pain.

The drive to model as closely as possible the clinical conditions in which pain occurs in humans has led to the development of animal models to reproduce the conditions for neuropathic pains associated with spinal or foraminal stenosis and disk herniations, many of which are considered critical to the development of chronic back pain. In these animal models, two L-shaped rods are placed unilaterally into the intravertebral foramin, one at L4 and the other at L5 (Hu and Xing 1998). The rods remain in place from 1 to 14 days, after which behavioral, electrophysiological, and anatomical studies are performed to document mechanical and thermal hypersensitivity and to elucidate the underlying causes of the pain. To what extent the pain that results from this condition reflects the compression and associated block of activity of subpopulations of afferent nerve fibers or whether there is an active inflammatory process that activates nerve fibers is a critical focus of study. In this regard it is of interest that the application of a variety of cytokines to the peripheral nerve (Sorkin et al. 1997) or even of autologous nucleus pulposus to the DRG of the rabbit (Cavanaugh et al. 1997) can recapitulate features of neuropathic pain.

Cancer Pain

As cancer pain is one of the most severe and most difficult pains to treat in humans, particularly in late stages of the disease, it is perhaps surprising that animal models of pain associated with cancer have only recently been developed. In part, the paucity of models reflects the difficulty of creating a reliable and reproducible condition. The last decade, however, has seen the development of such models in both rats and mice (for a review, Pacharinsak and Beitz 2008). Rather than studying the pain associated with the destruction of a particular organ, attention has focused on the pain that develops after metastasis of tumors to, for example, bone, which is among the most painful conditions. To this end, Mantyh and colleagues (Schwei et al. 1999) initially described a model that involved implanting osteolytic sarcoma cells in the femur of a mouse and sealing the femur to restrict tumor growth. Pathological studies as the tumor developed revealed characteristic osteoclast destruction of bone, presumably in the relatively acidic environment that promotes osteoclast function. Over time there was bone destruction concurrent with the development of a clear hypersensitivity to mechanical probing of the affected limb. Importantly, this model has proven very useful for the testing of novel pharmaceuticals for the treatment of pain associated with tumor metastasis to bone. Ongoing studies are directed at assessing the nature of the pathology that generates the pain. It was originally assumed that such cancer pains are largely inflammatory in nature, but animal studies indicate that there is a nerve injury-associated component as well. The peripheral nerve endings of fibers that innervate bone are unquestionably involved and these likely contribute to the mechanical hypersensitivity and ongoing pain that develop.

More recently, attention has turned to pains likely associated with the more traditional models of cancer that are used to study the biological basis for the generation and treatment of tumor development. For example, Lindsay and colleagues (2005) used a well-studied transgenic model of pancreatic cancer (produced by expression of the simian virus 40 large T antigen under control of the rat elastase-1 promoter) to monitor behavioral changes that might indicate ongoing pain. Interestingly, they found that when there were cellular changes characteristic of an inflammatory response, the mice did not manifest any behavior indicative of ongoing pain or hypersensitivity. A comparable magnitude of inflammatory changes in the skin would typically be associated with clear mechanical and thermal hypersensitivity. Signs of pain, including hunching and vocalization, eventually occurred at 16 weeks of age, at which point the pancreatic cancer was severe. Whether there is a masking of pain in the early stages of the disease remains to be determined, but this model illustrates that the mechanism(s) of development of the pains associated with different types of cancer are not the same and likely have multiple etiologies.

Spontaneous Pain

Most of the persistent pain models described above measure pain provoked by thermal, mechanical, or (less frequently) chemical stimuli. Many of these models are also presumed to be associated with ongoing, spontaneous pain, which frequently manifests as reduced activity. For example, in inflammatory visceral pain models, mice and rats with inflamed stomachs, bladders, or colons tend to sit quietly in their cages and do not explore in open field tests (although they do not become difficult to handle and they continue to eat and gain weight). Similarly, animals with inflamed or incised hindpaws commonly guard the paw by raising it above the floor and holding it in an unnatural posture. In tests these animals will not readily bear weight on the affected hindpaw until resolution of the insult. In both of the above examples, and in inflammatory models in general (e.g., joint, muscle, orofacial), the effects of the inflammation or incision are reversible and relatively short-lived (days to weeks). Whether ongoing pain at rest is present in these models is unknown. In analogous inflammatory and postsurgical circumstances in humans, pain at rest is either minimal or acceptable, but, as in these animal models, hypersensitivity and pain can be easily provoked by certain stimuli (e.g., forced movement, application of noxious stimuli).

In models of peripheral neuropathic pain, in which mechanical allodynia is present, nail growth and changes in hindpaw temperature (indicative of altered sympathetic efferent function) along with limb guarding are common. Cancer pain models are also associated with increasing discomfort and spontaneous pain as tumor burden increases. In both of these models, the effects of either nervous system insult or cancer are long-lasting (weeks to months) and minimally reversible; therefore, animals are generally euthanized according to humane endpoint principles.

Readers are urged to consult Chapter 5 for an extensive discussion of humane endpoints and Chapter 4 for an analysis of the ethical conflicts associated with research using persistent pain models.

REFERENCES

Abbott FV, Guy ER. 1995. Effects of morphine, pentobarbital and amphetamine on formalin-induced behaviours in infant rats: Sedation versus specific suppression of pain. Pain 62:303-312.

Al-Chaer ED, Kawasaki M, Pasricha PJ. 2000. A new model of chronic visceral hypersensitivity in adult rats induced by colon irritation during postnatal development. Gastroenterology 119(5):1276-1285.

Aley KO, Reichling DB, Levine JD. 1996. Vincristine hyperalgesia in the rat: A model of painful vincristine neuropathy in humans. Neuroscience 73(1):259-265.

Allen JW, Yaksh TL. 2004. Tissue injury models of persistent nociception in rats. Methods Mol Med 99:25-34.

Banik RK, Woo YC, Park SS, Brennan TJ. 2006. Strain and sex influence on pain sensitivity after plantar incision in the mouse. Anesthesiology 105(6):1246-1253.

Bendele A, McComb J, Gould T, McAbee T, Sennello G, Chlipala E, Guy M. 1999. Animal models of arthritis: Relevance to human disease. Toxicol Pathol 27(1):134-142.

Bennett GJ, Xie YK. 1988. A peripheral mononeuropathy in rat that produces disorders of pain sensation like those seen in man. Pain 33(1):87-107.

Berkley KJ, Wood E, Scofield SL, Little M. 1995. Behavioral responses to uterine or vaginal distension in the rat. Pain 61(1):121-131.

Berkley KJ, McAllister SL, Accius BE, Winnard KP. 2007. Endometriosis-induced vaginal hyperalgesia in the rat: Effect of estropause, ovariectomy, and estradiol replacement. Pain 132 Suppl 1:S150-S159.

Bishop T, Hewson DW, Yip PK, Fahey MS, Dawbarn D, Young AR, McMahon SB. 2007. Characterisation of ultraviolet-B-induced inflammation as a model of hyperalgesia in the rat. Pain 131(1-2):70-82.

Brand DD, Kang AH, Rosloniec EF. 2004. The mouse model of collagen-induced arthritis. Methods Mol Med 102:295-312.

Brennan TJ, Vandermeulen EP, Gebhart GF. 1996. Characterization of a rat model of incisional pain. Pain 64(3):493-501.

Browne KD, Iwata A, Putt ME, Smith DH. 2006. Chronic ibuprofen administration worsens cognitive outcome following traumatic brain injury in rats. Exp Neurol 201(2):301-307.

Burstein R, Yamamura H, Malick A, Strassman AM. 1998. Chemical stimulation of the intracranial dura induces enhanced responses to facial stimulation in brain stem trigeminal neurons. J Neurophysiol 79(2):964-982.

Burton MB, Gebhart GF. 1995. Effects of intracolonic acetic acid on responses to colorectal distension in the rat. Brain Res 672(1-2):77-82.

Caterina MJ, Leffler A, Malmberg AB, Martin WJ, Trafton J, Petersen-Zeitz KR, Koltzenburg M, Basbaum AI, Julius D. 2000. Impaired nociception and pain sensation in mice lacking the capsaicin receptor. Science 288(5464):306-13.

Cavanaugh JM, Ozaktay AC, Yamashita T, Avramov A, Getchell TV, King AI. 1997. Mechanisms of low back pain: A neurophysiologic and neuroanatomic study. Clin Orthop Relat Res (335):166-180.

Christianson JA, Gebhart GF. 2007. Assessment of colon sensitivity by luminal distension in mice. Nat Protoc 2(10):2624-2631.

Clavelou P, Dallel R, Orliaguet T, Woda A, Raboisson P. 1995. The orofacial formalin test in rats: Effects of different formalin concentrations. Pain 62(3):295-301.

Coderre TJ, Xanthos DN, Francis L, Bennett GJ. 2004. Chronic post-ischemia pain (CPIP): A novel animal model of complex regional pain syndrome-type I (CRPS-I; reflex sympathetic dystrophy) produced by prolonged hindpaw ischemia and reperfusion in the rat. Pain 112(1-2):94-105.

Coutinho SV, Meller ST, Gebhart GF. 1996. Intracolonic zymosan produces visceral hyperalgesia in the rat that is mediated by spinal NMDA and non-NMDA receptors. Brain Res 736(1-2):7-15.

Coutinho SV, Plotsky PM, Sablad M. 2002. Neonatal maternal separation alters stress-induced responses to viscerosomatic nociceptive stimuli in rat. Am J Physiol 282:G307-G16.

Decosterd I, Woolf CJ. 2000. Spared nerve injury: An animal model of persistent peripheral neuropathic pain. Pain 87(2):149-158.

Duarte AM, Pospisilova E, Reilly E, Mujenda F, Hamaya Y, Strichartz GR. 2005. Reduction of postincisional allodynia by subcutaneous bupivacaine: Findings with a new model in the hairy skin of the rat. Anesthesiology 103(1):113-125.

Dubuisson D, Dennis SG. 1977. The formalin test: A quantitative study of the analgesic effects of morphine, meperidine, and brain stem stimulation in rats and cats. Pain 4(2):161-174.

Ebersberger A, Handwerker HO, Reeh PW. 1999. Nociceptive neurons in the rat caudal trigeminal nucleus respond to blood plasma perfusion of the subarachnoid space: The involvement of complement. Pain 81(3):283-288.

Gebhart GF, Sengupta JN. 1996. Evaluation of visceral pain. In: Gaginella T, ed. Handbook of Methods in Gastrointestinal Pharmacology. Boca Raton, FL: CRC Press. Pp. 359-373.

Giamberardino MA, Valente R, de Bigontina P, Vecchiet L. 1995. Artificial ureteral calculosis in rats: Behavioural characterization of visceral pain episodes and their relationship with referred lumbar muscle hyperalgesia. Pain 61(3):459-469.

Griffiths MM, Cannon GW, Corsi T, Reese V, Kunzler K. 2007. Collagen-induced arthritis in rats. In Arthritis: Cope A, ed. Methods and Protocols, Vol. II. Humana Press. pp. 201-214.

Hartwig AC, Mathias SI, Law AS, Gebhart GF. 2003. Characterization and opioid modulation of inflammatory temporomandibular joint pain in the rat. J Oral Maxillofac Surg 61(11):1302-1309.

Hong Y, Abbott FV. 1994. Behavioural effects of intraplantar injection of inflammatory mediators in the rat. Neuroscience 63(3):827-836.

Honoré P, Buritova J, Besson JM. 1995. Aspirin and acetaminophen reduced both Fos expression in rat lumbar spinal cord and inflammatory signs produced by carrageenin inflammation. Pain 63(3):365-75

Hu SJ, Xing JL. 1998. An experimental model for chronic compression of dorsal root ganglion produced by intervertebral foramen stenosis in the rat. Pain 77(1):15-23.

Hunskaar S, Hole K. 1987. The formalin test in mice: Dissociation between inflammatory and non-inflammatory pain. Pain 30(1):103-114.

Iadarola MJ, Brady LS, Draisci G, Dubner R. 1988. Enhancement of dynorphin gene expression in spinal cord following experimental inflammation: Stimulus specificity, behavioral parameters and opioid receptor binding. Pain 35(3):313-326.

Jones RC, Otsuka E, Wagstrom E, Jensen CS, Price MP, Gebhart GF. 2007. Short-term sensitization of colon mechanoreceptors is associated with long-term hypersensitivity to colon distention in the mouse. Gastroenterology 133(1):184-194.

Kamp EH, Jones RC 3rd, Tillman SR, Gebhart GF. 2003. Quantitative assessment and characterization of visceral nociception and hyperalgesia in mice. Am J Physiol Gastrointest Liver Physiol 284(3):G434-G444.

Kaneko M, Mestre C, Sanchez EH, Hammond DL. 2000. Intrathecally administered gabapentin inhibits formalin-evoked nociception and the expression of Fos-like immunoreactivity in the spinal cord of the rat. J Pharmacol Exp Therap 292:743-751.

Kim SH, Chung JM. 1992. An experimental model for peripheral neuropathy produced by segmental spinal nerve ligation in the rat. Pain 50(3):355-363.

Lamb K, Kang YM, Gebhart GF, Bielefeldt K. 2003. Gastric inflammation triggers hypersensitivity to acid in awake rats. Gastroenterology 125(5):1410-1418.

Lantéri-Minet M, Bon K, de Pommery J, Michiels JF, Menétrey D. 1995. Cyclophosphamide cystitis as a model of visceral pain in rats: Model elaboration and spinal structures involved as revealed by the expression of c-Fos and Krox-24 proteins. Exp Brain Res 105(2):220-232.

Lariviere WR, Melzack R. 1996. The bee venom test: A new tonic-pain test. Pain 66(2-3): 271-277.

Law AS, Baumgardner KR, Meller ST, Gebhart GF. 1999. Localization and changes in NADPH-diaphorase reactivity and nitric oxide synthase immunoreactivity in rat pulp following tooth preparation. J Dent Res 78(10):1585-1595.

LeBars D, Gozuriu M, Cadden SW. 2001. Animal models of nociception. Pharmacol Rev 53:597-652.

Lindsay TH, Jonas BM, Sevcik MA, Kubota K, Halvorson KG, Ghilardi JR, Kuskowski MA, Stelow EB, Mukherjee P, Gendler SJ, Wong GY, Mantyh PW. 2005. Pancreatic cancer pain and its correlation with changes in tumor vasculature, macrophage infiltration, neuronal innervation, body weight and disease progression. Pain 119(1-3):233-246.

Malmberg AB, Basbaum AI. 1998. Partial sciatic nerve injury in the mouse as a model of neuropathic pain: Behavioral and neuroanatomical correlates. Pain 76(1-2):215-222.

Meller ST, Gebhart GF. 1997. Intraplantar zymosan as a reliable, quantifiable model of thermal and mechanical hyperalgesia in the rat. Eur J Pain 1(1):43-52.

Moran EL, Bogoch ER. 1999. Animal models of rheumatoid arthritis. In: An YH, Friedman, RJ, eds. Animal Models in Orthopaedic Research. CRC Press, pp. 369-393.

Morgan JR, Gebhart GF. 2008. Characterization of a model of chronic orofacial hyperalgesia in the rat: Contribution of NA(V) 1.8. J Pain 9(6):522-531.

Morris GP, Beck PL, Herridge MS, Depew WT, Szewczuk MR, Wallace JL. 1989. Hapten-induced model of chronic inflammation and ulceration in the rat colon. Gastroenterology 96(3):795-803.

Ness TJ, Elhefni H. 2004. Reliable visceromotor responses are evoked by noxious bladder distention in mice. J Urol 171(4):1704-1708.

Ness TJ, Gebhart GF. 1990. Visceral pain: A review of experimental studies. Pain 41(2): 167-234.

Ness TJ, Lewis-Sides A, Castroman P. 2001. Characterization of pressor and visceromotor reflex responses to bladder distention in rats: Sources of variability and effect of analgesics. J Urol 165(3):968-974.

Neugebauer V, Han JS, Adwanikar H, Fu Y, Ji G. 2007. Techniques for assessing knee joint pain in arthritis. Mol Pain 3:8.

Nozaki-Taguchi N, Yaksh TL. 1998. A novel model of primary and secondary hyperalgesia after mild thermal injury in the rat. Neurosci Lett 254(1):25-28.

Olechowski CJ, Truong JJ, Kerr BJ. 2009. Neuropathic pain behaviours in a chronic-relapsing model of experimental autoimmune encephalomyelitis (EAE). Pain 141(1-2):156-64.

Ozaki N, Bielefeldt K, Sengupta JN, Gebhart GF. 2002. Models of gastric hyperalgesia in the rat. Am J Physiol Gastrointest Liver Physiol 283(3):G666-G676.

Pacharinsak C, Beitz A. 2008. Animal models of cancer pain. Comp Med 58(3):220-233.

Polomano RC, Mannes AJ, Clark US, Bennett GJ. 2001. A painful peripheral neuropathy in the rat produced by the chemotherapeutic drug, paclitaxel. Pain 94(3):293-304.

Radhakrishnan R, Moore SA, Sluka KA. 2003. Unilateral carrageenan injection into muscle or joint induces chronic bilateral hyperalgesia in rats. Pain 104(3):567-577.

Rakieten N, Rakieten ML, Nadkarni MV. 1963. Studies on the diabetogenic action of streptozotocin (NSC-37917). Cancer Chemother Rep 29:91-98.

Randich A, Uzzell T, DeBerry JJ, Ness TJ. 2006a. Neonatal urinary bladder inflammation produces adult bladder hypersensitivity. J Pain 7(7):469-479.

Randich A, Uzzell T, Cannon R, Ness TJ. 2006b. Inflammation and enhanced nociceptive responses to bladder distension produced by intravesical zymosan in the rat. BMC Urol 6:2.

Saddi G-M, Abbott FV. 2000. The formalin test in the mouse: A parametric analysis of scoring properties. Pain 89:53-63.

Sadzot-Delvaux C, Merville-Louis MP, Delree P, Marc P, Piette J, Moonen G, Rentier B. 1990. An in vivo model of varicella-zoster virus latent infection of dorsal root ganglia. J Neurosci Res 26(1):83-89.

Schwei MJ, Honore P, Rogers SD, Salak-Johnson JL, Finke MP, Ramnaraine ML, Clohisy DR, Mantyh PW. 1999. Neurochemical and cellular reorganization of the spinal cord in a murine model of bone cancer pain. J Neurosci 19(24):10886-10897.

Seltzer Z, Dubner R, Shir Y. 1990. A novel behavioral model of neuropathic pain disorders produced in rats by partial sciatic nerve injury. Pain 43(2):205-218.

Shields SD, Eckert WA 3rd, Basbaum AI. 2003. Spared nerve injury model of neuropathic pain in the mouse: A behavioral and anatomic analysis. J Pain 4(8):465-470.

Sluka KA, Westlund KN. 1993. Behavioral and immunohistochemical changes in an experimental arthritis model in rats. Pain 55(3):367-377.

Sluka KA, Kalra A, Moore SA. 2001. Unilateral intramuscular injections of acidic saline produce a bilateral, long-lasting hyperalgesia. Muscle Nerve 24(1):37-46.

Smith SB, Crager SE, Mogil JS. 2004. Paclitaxel-induced neuropathic hypersensitivity in mice: Responses in 10 inbred mouse strains. Life Sci 74(21):2593-2604.

Sorkin LS, Xiao WH, Wagner R, Myers RR. 1997. Tumour necrosis factor-alpha induces ectopic activity in nociceptive primary afferent fibres. Neuroscience 81(1):255-262.

Tsuboi Y, Takeda M, Tanimoto T, Ikeda M, Matsumoto S, Kitagawa J, Teramoto K, Simizu K, Yamazaki Y, Shima A, Ren K, Iwata K. 2004. Alteration of the second branch of the trigeminal nerve activity following inferior alveolar nerve transection in rats. Pain 111(3):323-334.

Van den Berg WB, Joosten LAB, van Lent PLEM. 2007. Murine antigen-induced arthritis. In: Cope A, ed. Arthritis: Methods and Protocols, Vol. II. Humana Press. pp. 243-253.

Vera-Portocarrero LP, Lu Y, Westlund KN. 2003. Nociception in persistent pancreatitis in rats: Effects of morphine and neuropeptide alterations. Anesthesiology 98(2):474-484.

Vilensky JA, O'Connor BL, Brandt KD, Dunn EA, Rogers PI, DeLong CA. 1994. Serial kinematic analysis of the unstable knee after transection of the anterior cruciate ligament: Temporal and angular changes in a canine model of osteoarthritis. J Orthop Res 12(2):229-237.

Vos BP, Strassman AM, Maciewicz RJ. 1994. Behavioral evidence of trigeminal neuropathic pain following chronic constriction injury to the rat's infraorbital nerve. J Neurosci 14(5 Pt 1):2708-2723.

Wallace VC, Blackbeard J, Segerdahl AR, Hasnie F, Pheby T, McMahon SB, Rice AS. 2007. Characterization of rodent models of HIV-gp120 and anti-retroviral-associated neuropathic pain. Brain 130(Pt 10):2688-2702.

Wesselmann U, Czakanski PP, Affaitati G, Giamberardino MA. 1998. Uterine inflammation as a noxious visceral stimulus: Behavioral characterization in the rat. Neurosci Lett 246(2):73-76.

Wuarin-Bierman L, Zahnd GR, Kaufmann F, Burcklen L, Adler J. 1987. Hyperalgesia in spontaneous and experimental animal models of diabetic neuropathy. Diabetologia 30(8): 653-658.

Yezierski RP, Liu S, Ruenes GL, Kajander KJ, Brewer KL. 1998. Excitotoxic spinal cord injury: Behavioral and morphological characteristics of a central pain model. Pain 75(1): 141-155.

Young W. 2002. Spinal cord contusion models. Prog Brain Res 137:231-255.

APPENDIX

B

US Regulations and Guidelines Regarding Recognition and Alleviation of Pain in Laboratory Animals

The requirement or recommendation to consider the recognition and alleviation of pain in laboratory animals when conducting research in the United States is constituted in federal law, regulations, and guidelines, enforced by the US Public Health Service Policy, and promulgated by various professional organizations as outlined below.

LEGAL REQUIREMENTS AND AGENCY GUIDELINES

US Animal Welfare Act

The primary federal regulation concerning the care and use of laboratory animals in the United States is the Animal Welfare Act (AWA; Public Law 89-544 as amended, 7 USC Ch. 54). The AWA is implemented through the Animal Welfare Act Regulations, published in the Code of Federal Regulations (CFR), Title 9, Chapter 1, Subchapter A. The Act covers pets and warm-blooded animals used for research, testing, and exhibition purposes, but does not protect a number of animal species; for example, it specifically excludes rats of the genus *Rattus*, mice of the genus *Mus*, and birds bred for use in research.

The Animal Welfare Regulations consider painful procedures and methods to alleviate pain in several sections:

- §2.31(a), (d): Registered research institutions must have an institutional animal care and use committee (IACUC) that reviews and approves all procedures conducted using laboratory animals.

- §2.31(d)(i): "Procedures involving animals will avoid or minimize discomfort, distress or pain to animals."
- §2.31(d)(ii): "The principal investigator has considered alternatives to procedures that may cause more than momentary or slight pain."
- §2.31(e): "A proposal to conduct an activity involving animals . . . must contain . . . a description of procedures designed to assure that discomfort and pain to animals will be limited to that which is unavoidable for the conduct of scientifically valuable research including provision for the use of analgesic, anesthetic, and tranquilizing drugs where indicated and appropriate to minimize discomfort and pain to animals."
- §2.33(a): "Each research facility shall have an attending veterinarian who shall provide adequate veterinary care to animals in compliance with this section."
- §2.33(b)(4): The attending veterinarian shall provide "guidance to principal investigators and other personnel involved in the care and use of animals regarding handling, immobilization, anesthesia, tranquilization, and euthanasia."

The United States Department of Agriculture (USDA) is responsible for the administration and enforcement of this act through its Animal and Plant Health Inspection Service (APHIS).

USDA Policies

The USDA through APHIS periodically issues and updates policies to clarify the provisions of the Animal Welfare Regulations and provide improved guidance to USDA personnel who inspect the regulated research programs. Two USDA policies address the requirement to recognize the potential for pain in association with research activities.

Policy #11—"Painful Procedures"

Policy #11 (dated April 14, 1997) defines a painful procedure as "any procedure that would reasonably be expected to cause more than slight or momentary pain and/or distress in a human being to which the procedure is applied" and requires the IACUC to ensure that investigators have considered appropriate alternatives to such procedures. The policy lists examples of procedures that are likely to cause more than momentary or slight pain, including but not limited to terminal surgery (alleviated by anesthesia), use of complete Freund's adjuvant (depending on the product, procedure, and species), and ocular and skin irritancy testing. The policy further states the expectation that animals exhibiting signs of pain or discomfort will receive appropriate pain relief unless justified scientifically, in writing, and approved by the IACUC. Policy #11 also requires the reporting of animals subjected to procedures that may cause pain and its alleviation through the use of anesthetics, analgesics, sedatives, and/or tranquilizers, as well as the separate reporting of animals subjected to such procedures in which pain-relieving agents were not administered, for IACUC-approved research requirements.

Policy #12—"Considerations of Alternatives to Painful/Distressful Procedures"

This policy (dated June 21, 2000) provides guidance for the AWA requirement that principal investigators consider alternatives to painful procedures. Such alternatives should include some aspect of replacement, reduction, or refinement of animal use to minimize animal pain consistent with research goals. For procedures that may cause pain, the policy states that "any proposed animal activity, or significant changes to an ongoing animal activity, must include: a description of procedures or methods designed to assure that discomfort and pain to animals will be limited to that which is unavoidable in the conduct of scientifically valuable research, and that analgesic, anesthetic, and tranquilizing drugs will be used where indicated and appropriate to minimize discomfort and pain to animals." The policy also requires that proposed animal use include "a written description of the methods and sources used to consider alternatives to procedures that may cause more than momentary or slight pain to animals."

Health Research Extension Act

The Health Research Extension Act (Public Law 99-158, November 20, 1985, "Animals in Research") provides the statutory mandate for the Public Health Service Policy on Humane Care and Use of Laboratory Animals (OLAW 2002 reprint of PHS Policy; preface). The Act mandates that "the Secretary [of the US Department of Health and Human Services], acting through the Director of NIH, shall establish guidelines for the following . . . [in procedures that may cause pain]: the proper treatment of animals while being used in research . . . shall require the appropriate use of tranquilizers, analgesics, anesthetics, paralytics, and euthanasia for animals." The PHS Policy (see below) defines procedures to implement this mandate.

Public Health Service Policy on Humane Care and Use of Laboratory Animals

The Public Health Service Policy on Humane Care and Use of Laboratory Animals (PHS Policy) (DHHS 2002) was introduced in 1973 and has been revised multiple times (most recently in 2002). The Policy applies to all institutions that use animals in research that is supported by any component of the PHS (e.g., NIH, CDC, FDA) and it "requires institutions to establish and maintain proper measures to ensure the appropriate care and use of all animals involved in research, training, and biological testing." While the PHS Policy mandates compliance with the AWA and AWA Regulations, it uses a broader definition of an animal: "any live, vertebrate animal used or intended for use in research, training, experimentation, or biological testing." Further, the Policy endorses the US Government Principles for the Utilization and Care of Vertebrate Animals Used in Testing, Research, and Training (see below) and requires institutions to base their animal care and use programs on the National Research Council's *Guide for the Care and Use of Laboratory Animals* (NRC 1996).

The PHS Policy defines procedures for submission of the Animal Welfare Assurance statement, which is required of all institutions that conduct PHS-funded research, training, or testing with animals. For potentially painful procedures on animals, the PHS policy requires the IACUC to determine that "procedures with animals will avoid or minimize discomfort, distress, and pain to the animals, consistent with sound research design; procedures that may cause more than momentary or slight pain or distress to animals will be performed with appropriate sedation, analgesia, or anesthesia, unless the procedure is justified for scientific reasons in writing by the investigator; and animals that would otherwise experience severe or chronic pain or distress that cannot be relieved will be painlessly killed at the end of the procedure or, if appropriate, during the procedure." The PHS Policy further states that "methods of euthanasia used will be consistent with the recommendations of the American Veterinary Medical Association" (AVMA 2007). Additionally, with respect to potentially painful procedures, the PHS Policy requires applications and proposals for PHS awards to include "a description of procedures designed to assure that discomfort and injury to animals will be limited to that which is unavoidable in the conduct of scientifically valuable research, and that analgesic, anesthetic, and tranquilizing drugs will be used where indicated and appropriate to minimize discomfort and pain to animals."

The NIH Office of Laboratory Animal Welfare (OLAW) is responsible for administering the PHS Policy.

US Government Principles for the Utilization and Care of Vertebrate Animals Used in Testing, Research, and Training

The US Government Principles for the Utilization and Care of Vertebrate Animals Used in Testing, Research, and Training (US Government Principles) were drafted in 1985 by the Interagency Research Animal Committee (IRAC 1985). The document addresses compliance with federal laws, policies, and guidelines and establishes overarching principles to consider when using animals in research, testing, and training. Principles 4, 5, and 6 relate to the potential to cause pain in laboratory animals.

- Principle #4: "Proper use of animals, including the avoidance or minimization of discomfort, distress, and pain when consistent with sound scientific practices, is imperative. Unless the contrary is established, investigators should consider that procedures that cause pain or distress in human beings may cause pain or distress in other animals."
- Principle #5: "Procedures with animals that may cause more than momentary or slight pain or distress should be performed under appropriate sedation, analgesia, or anesthesia. Surgical or other painful procedures should not be performed on unanesthetized animals paralyzed by chemical agents."
- Principle #6: "Animals that would otherwise suffer severe or chronic pain or distress that cannot be relieved should be painlessly killed at the end of the procedure, or if appropriate, during the procedure."

Guide for the Care and Use of Laboratory Animals

The recommendations and guidelines of the *Guide for the Care and Use of Laboratory Animals* (7th ed.; NRC 1996; the *Guide*) were drafted by a committee of the National Research Council's Institute for Laboratory Animal Research to promote the humane care and use of laboratory animals. The *Guide* emphasizes the application of performance standards and professional judgment and encourages users and institutions to achieve excellent standards of animal care and use by determining how best to achieve these goals within the scope and capabilities of the particular institution. The *Guide* also endorses the responsibilities of investigators as stated in the US Government Principles (IRAC 1985; outlined above). Both the PHS Policy and the Association for Assessment and Accreditation of Laboratory Animal Care International (AAALAC International) require institutions to base their programs of animal care and use on the recommendations detailed in the *Guide*.

The *Guide* calls for the establishment of an IACUC, which must ensure the appropriate application of sedation, analgesia, and anesthesia when reviewing protocols (p. 9), and notes that "ethical, humane, and scientific considerations sometimes require the use of sedatives, analgesics, or anesthetics in animals" (p. 12). The *Guide* (p. 64) also devotes a section to the consideration of pain, analgesia, and anesthesia, and states that "an integral component of veterinary medical care is prevention or alleviation of pain associated with procedural and surgical protocols." Although recognizing such pain is complex and can be challenging, the *Guide* indicates that the ability to experience and respond to pain is widespread in the animal kingdom. The *Guide* therefore stipulates that the proper use of anesthetics and analgesics in research animals is an ethical and scientific imperative and that in general, unless the contrary is known or established, it should be assumed that procedures that cause pain in humans also cause pain in animals.

OTHER RELEVANT GUIDELINES AND STATEMENTS

Association for the Assessment and Accreditation of Laboratory Animal Care International

The Association for the Assessment and Accreditation of Laboratory Animal Care (AAALAC) International is a private, nonprofit organization that promotes the humane care and use of laboratory animals through a program of voluntary assessment and accreditation. AAALAC International does not itself define standards but rather uses the *Guide* as its primary assessment resource along with other peer-reviewed reference standards. Additionally, when conducting assessments of accredited programs, AAALAC International requires that institutions comply with applicable principles, regulations, standards, policies, and guidelines concerning pain in laboratory animals.

American College of Laboratory Animal Medicine

The American College of Laboratory Animal Medicine (ACLAM) is the professional organization of veterinarians who have completed the requirements for board certification as specialists in laboratory animal medicine. ACLAM has issued position statements regarding Adequate Veterinary Care and Pain and Distress in Laboratory Animals.

ACLAM Report on Adequate Veterinary Care

The *Report on Adequate Veterinary Care* (ACLAM 1996) details the expectations and requirements of an institution's veterinary care program, including the expectation that the veterinarian will have the authority to ensure the proper use of anesthetics, analgesics, tranquilizers, and methods of euthanasia. The report further states that "written guidelines regarding the selection and use of anesthetics, analgesics and tranquilizing drugs and euthanasia practices for all species used must be provided and periodically reviewed by the veterinarian." Additionally, "the veterinarian must have the responsibility and authority to assure that handling, restraint, anesthesia, analgesia and euthanasia are administered as required to relieve pain and suffering in research animals, provided such intervention is not specifically precluded in protocols reviewed and approved by the IACUC. The veterinarian must exercise good professional judgment to select the most appropriate pharmacologic agent(s) and methods to relieve animal pain or distress in order to assure humane treatment of animals, while avoiding undue interference with goals of the experiment."

The ACLAM Position Statement on Pain and Distress in Laboratory Animals

The ACLAM Position Statement on Pain and Distress in Laboratory Animals details the expectations of the College concerning pain in laboratory animals (ACLAM 2001):

> Procedures expected to cause more than slight or momentary pain (e.g., pain in excess of a needle prick or injection) require the appropriate use of pain-relieving measures unless scientifically justified in an approved animal care and use protocol. Requests for exceptions to the use of analgesics, tranquilizers, anesthetics or non-chemical means of providing relief from pain and/or distress must be scientifically justified by the Principal Investigator and approved by the Institutional Animal Care and Use Committee (IACUC) prior to initiation of the protocol. Paramount in the decision to provide relief from pain and distress is the professional judgment of a trained laboratory animal veterinarian. The *Guide for the Care and Use of Laboratory Animals* (NRC 1996) and the Animal Welfare Act emphasize the vital role of the veterinarian in this process—the attending veterinarian, or his/her designee, should recommend the pain- or distress-relieving measure or agent, dose, frequency, and duration of administration according to his/her professional judgment and clinical assessment of the research subject(s). Thus, veterinary participation is needed in the planning phase of those experiments with the potential to produce pain or distress and in the ongoing review of the animal's condition. Consideration should be given to preventing pain or distress by using preemptive measures whenever possible. While the animal care and use protocol must provide information on types of pain- and distress-relieving medications and treatments intended to be used, the veterinarian's clinical assessment and judgment regarding what is in the best interest of the animal should be given overriding precedence.

REFERENCES

ACLAM (American College of Laboratory Animal Medicine). 1996. ACLAM Report on Adequate Veterinary Care. Available at www.aclam.org/education/guidelines/position.html. Accessed June 9, 2008.

ACLAM. 2001. ACLAM Position Statement on Pain and Distress in Laboratory Animals. Available at www.aclam.org/education/guidelines/position_pain-distress.html. Accessed June 9, 2008.

AVMA (American Veterinary Medical Association). 2007. AVMA Guidelines on Euthanasia. Available at www.avma.org/issues/animal_welfare/euthanasia.pdf. Accessed June 9, 2008.

AWA (Animal Welfare Act). 1990. Animal Welfare Act. Available at www.nal.usda.gov/awic/legislat/awa.htm. Accessed June 9, 2008.

DHHS/NIH/OLAW (Department of Health and Human Services, National Institutes of Health, Office of Laboratory Animal Welfare). 2002. Public Health Service Policy on Humane Care and Use of Laboratory Animals. Available at http://grants.nih.gov/grants/olaw/references/phspol.htm. Accessed June 9, 2008.

IRAC (Interagency Research Animal Committee). 1985. The U.S. Government Principles for the Utilization and Care of Vertebrate Animals Used in Testing, Research, and Training. Federal Register Vol. 50, No. 97 (May 20, 1985). Available at http://grants.nih.gov/grants/olaw/references/phspol.htm#USGovPrinciples. Accessed June 9, 2008.

NRC (National Research Council). 1996. Guide for the Care and Use of Laboratory Animals. Washington: National Academy Press.

USDA (US Department of Agriculture). 2005a. 9 CFR 2.31. (Title 9, Volume 1, Part 2.31): Institutional Animal Care and Use Committee. Available at www.aphis.usda.gov/animal_welfare/downloads/awr.9cfr2.31.txt. Accessed January 5, 2009.

USDA. 2005b. 9 CFR 2.33. (Title 9, Volume 1, Part 2.33): Attending Veterinarian and Adequate Veterinary Care. Available at www.aphis.usda.gov/animal_welfare/downloads/awr.9cfr2.33.txt. Accessed January 5, 2009.

USDA-APHIS (USDA-Animal and Plant Health Inspection Service). 1997. APHIS Policy #11, "Painful Procedures" (dated: April 14, 1997). Available at www.aphis.usda.gov/animal_welfare/downloads/policy/policy11.pdf. Accessed June 9, 2008.

USDA-APHIS. 1997. APHIS Policy #12, "Considerations of Alternatives to Painful/Distressful Procedures" (dated: June 21, 2000). Available at www.aphis.usda.gov/animal_welfare/downloads/policy/policy12.pdf. Accessed June 9, 2008.

ERRATUM

Recognition and Alleviation of Pain in Laboratory Animals
ISBN-13: 978-0-309-12834-6; ISBN-10: 0-309-12834-X
The National Academies Press, Washington, D.C.

The following section was inadvertently omitted from the published book.

APPENDIX C

About the Authors

Gerald F. Gebhart (*Chair*), PhD, is Professor and Director of the Center for Pain Research at the University of Pittsburgh. He has more than three decades of experience in pain research that has focused on endogenous systems of pain control and mechanisms of hypersensitivity, most recently visceral hypersensitivity. Dr. Gebhart has developed widely used animal models for the study of mechanisms of postoperative, incisional, and visceral pain (stomach and colon). He has served on the ILAR Council, as editor of the *ILAR Journal,* and on the ILAR committees that produced *Recognition and Alleviation of Pain and Distress in Laboratory Animals* (1992) and the *Guide for the Care and Use of Laboratory Animals* (1996). He is a Past President of the American Pain Society, current Editor in Chief of the Society's *Journal of Pain,* and President (2008-2011) of the International Association for the Study of Pain.

Allan I. Basbaum, PhD, FRS, IOM, is Professor and Chair of the Department of Anatomy and a member of the W.M. Keck Foundation Center for Integrative Neurosciences at the University of California San Francisco. He has studied the peripheral and central nervous system mechanisms that underlie the generation and control of pain for over four decades. A major component of his research involves behavioral analysis of animals, including responses to peripheral stimulation in the setting of tissue or nerve injury. His laboratory uses a variety of injury conditions that model clinical pain states, so that novel therapeutic targets for the control of pain may be identified. Assessment and measurement of pain behavior are thus critical elements of the work performed in his laboratory. He is Editor in Chief of *Pain*, the journal of the International Association for the Study of Pain.

Stephanie J. Bird, PhD, is a laboratory-trained neuroscientist whose current research interests focus on ethical issues associated with scientific research, especially in the area of neuroscience. She is co-Editor in Chief of the journal *Science and Engineering Ethics*. As Special Assistant to the Provost and Vice President for Research at the Massachusetts Institute of Technology from 1992 to 2003, Dr. Bird worked on the development of educational programs that addressed ethical issues in science and engineering, research practice, and professional responsibilities. Dr. Bird is an active member of the Society for Neuroscience and former Chair of its Social Issues Committee (2003-2005). She is also an active member and Fellow of the American Association for the Advancement of Science (AAAS) and has been Secretary of its Societal Impacts of Science and Engineering Section since 1995. Dr. Bird has been a member of the Tufts University Animal Care and Use Committee since 1991.

Paul Flecknell, MA, VetMB, PhD, is Professor at the Medical School of Newcastle University. He is a Diplomate of the European Colleges of Veterinary Anaesthesia and Analgesia and Laboratory Animal Medicine and a Diplomate of the UK Royal College of Veterinary Medicine in Laboratory Animal Science. He is also a veterinarian and has a PhD in physiology. He has done research in the area of animal anesthesia and analgesia for over 25 years and has published extensively in these fields. He also serves as the clinical veterinarian at a large multispecies research animal unit and is actively involved in implementing pain assessment and alleviation techniques in a range of species. He teaches pain management to a number of different groups on a regular basis.

Lyndon J. Goodly, DVM, MS, is Associate Vice Chancellor for Research at the University of Illinois Urbana-Champaign. As a Diplomate of the American College of Laboratory Animal Medicine with over 16 years of experience in the field, he has worked with a vast array of animals including amphibians, cats, dogs, fish, nonhuman primates, rodents, swine, and other agricultural species. He has served as an ad hoc member of two NIH Special Emphasis Panels and as a voting member of a number of institutional animal care and use committees.

Alicia Z. Karas, MS, DVM, is Assistant Professor in Tufts University's Cummings School of Veterinary Medicine. She teaches anesthesiology and pain medicine and works extensively with researchers, IACUCs, and laboratory animal organizations to promote and lecture on best practices of current veterinary pain medicine. She was a member of the school's IACUC and has been its Vice Chair since 1999. She is also a member of the Board of Directors of the International Veterinary Academy of Pain Management. Her research areas include methods of assessment and treatment of pain in mice, rabbits, dogs, and goats, improved methods of handling laboratory animals, and humane endpoints. She is on the editorial board for *Lab Animal* magazine and is an editor and author of the 2008 version of the ACLAM text *Anesthesia and Analgesia of Laboratory Animals*.

Stephen T. Kelley, DVM, MS, is Clinical Associate Professor at the University of Washington and a retired Supervisory Veterinarian and Head of Veterinary Medicine and Surgery at the Washington National Primate Research Center. As a Diplomate of the American College of Laboratory Animal Medicine, he has over 33 years of experience working with nonhuman primates (both Old and New World species) in clinical and research settings. Additionally, Dr. Kelley has served as a member of the Council on Accreditation of the Association for Assessment and Accreditation of Laboratory Animal Care (AAALAC) International since 1998.

Jane Lacher, DVM, is Clinical Veterinarian for the Dow Chemical Company Toxicology and Environmental Research and Consulting Laboratory, where she is responsible for the care and welfare of a variety of animal species in studies of chronic oncogenicity, metabolism, immunotoxicology, neurotoxicology, and respiratory, acute, genetic, reproductive, and developmental toxicology. Dr. Lacher dialogues with and advises coworkers on humane practices and endpoints involving animals in toxicology studies and is a member of the Dow Chemical Company Animal Welfare Opportunity Team responsible for ensuring a corporate commitment to animal welfare, both within the corporation and for studies at contract research organizations.

Georgia Mason, PhD, is Canada Research Chair in Animal Welfare and a member of the IACUC at the University of Guelph. Her main research interest is the chronic effects of standard housing on brain, behavior, and welfare. She is particularly interested in the use of behavioral measures (e.g., preference/avoidance; abnormal activities such as stereotypy) in objective welfare assessment. Her laboratory animal welfare projects include studies on the effects of early enrichment on later welfare in mice; of different enrichments on alopecia, aggression, and corticosteroid excretion in rhesus macaques; of different cage-cleaning regimes on rat and mouse welfare (in collaboration with Harlan UK); and of weaning age on mouse anxiety. She also studies the use of chromodacryorrhea and corticosterone from single micturations in assessing acute stress in the rat.

Lynne U. Sneddon, PhD, is Lecturer at the University of Liverpool. Her current research program examines pain, fear, and stress in fish using fMRI and other techniques in neuroanatomy, neurophysiology, genomics, whole animal physiology, and behavior. She first identified nociceptors in fish and embarked on projects aimed at understanding the importance of the nociceptive experience to fish and how to alleviate their pain by examining a number of analgesics. She was part of the working group of the Council of Europe's Farmed Fish Welfare guidelines endorsed in June 2006. She is a member of the European Food Safety Association's working group on farmed fish welfare and of the Association for the Study of Animal Behavior (ASAB) ethical committee. She has served as an advisor to numerous societies (including the Canadian Care Council) on their guidelines regarding fish.

Sulpicio G. Soriano, MD, MSEd, FAAP, is the Children's Hospital Boston Endowed Chair in Pediatric Neuroanesthesia and Associate Professor of Anesthesia at Harvard Medical School. He has been involved in anesthesia-related investigations in laboratory animals for 20 years and is recently studying the effects of anesthetic drugs on inflammation and the developing central nervous system. In his clinical role as a pediatric neuroanesthesiologist, he advocates the humane use of anesthesia and analgesia in animal research.

Consultant

Heidi L. Shafford, DVM, PhD, is a Consultant in anesthesia and pain management for research facilities and veterinary teams. She is a Diplomate of the American College of Veterinary Anesthesiologists. For over 10 years Dr. Shafford has been involved in studying the physiologic and behavioral effects of pain and analgesics in a variety of laboratory animal models. Concurrently, she assisted IACUCs, investigators, and veterinary staff to establish protocols for preventing and treating pain. Dr. Shafford owns and operates Veterinary Anesthesia Specialists, LLC, in Portland, Oregon. She regularly provides training related to anesthetic and analgesic practices for industry, academic, private, and professional organizations nationwide.

Index

B

C

D

E

I

J

K

L

M

R

S

T

U

V

W

X

Z